GLOBAL COMPANIES, LOCAL INNOVATIONS

T0230568

Global Companies, Local Innovations

Why the Engineering Aspects of Innovation Making Require Co-location

YASUYUKI MOTOYAMA
Kauffman Foundation, USA

Routledge
Taylor & Francis Group

LONDON AND NEW YORK

First published 2012 by Ashgate Publishing

Published 2016 by Routledge
2 Park Square, Milton Park, Abingdon, Oxfordshire OX14 4RN
711 Third Avenue, New York, NY 10017, USA

First issued in paperback 2016

Routledge is an imprint of the Taylor & Francis Group, an informa business

British Library Cataloguing in Publication Data
Motoyama, Yasuyuki.
Global companies, local innovations : why the engineering aspects of innovation making require co-location. – (Ashgate economic geography series)
1. International business enterprises–Location. 2. Industrial location–Effect of labor market on. 3. Product design. 4. New products–International cooperation. 5. Toyota Jidosha Hanbai Kabushiki Kaisha. 6. Soni Kabushiki Kaisha. 7. Kyanon Kabushiki Kaisha.
I. Title II. Series
338.8'81-dc23

Library of Congress Cataloging-in-Publication Data
Motoyama, Yasuyuki.
Global companies, local innovations : why the engineering aspects of innovation making require co-location / by Yasuyuki Motoyama.
  p. cm. – (Ashgate economic geography series)
Includes index.
ISBN 978-1-4094-2146-7 (hardback : alk. paper) 1.

International business enterprises. 2. Industries–Effect of technological innovations on. 3. Commercial geography. I. Title.
HD2756.M68 2012
338'.064–dc23

2012012007

ISBN 13: 978-1-138-27140-1 (pbk)
ISBN 13: 978-1-4094-2146-7 (hbk)

Contents

List of Figures

List of Tables

List of Abbreviations

ADR	American Depository Receipt
AV	Audio Visual
CAD	Computer Aided Design
CAE	Computer Aided Engineering
CEO	Chief Executive Officer
COO	Chief Operating Officer
CPU	Central Processing Unit
DEC	Digital Equipment Corporation
DRAM	Dynamic Random Access Memory
ECU	Electric Control Unit
HDD	Hard Disc Drive
HP	Hewlett-Packard
IBM	International Business Machine
IPO	Initial Public Offering
LCD	Liquid Crystal Display
LSI	Large-Scale Integrator
MNC	Multinational Corporation
MPG	Miles per Gallon
OEM	Original Equipment Manufacturing
OS	Operation System
RE	Resident Engineer
RQ	Research Question
SE	Simultaneous Engineering
SEC	Securities and Exchange Commission
TFP	Total Factor Productivity
VAIO	Video Audio Integrated Operation
VVTi	Variable Valve Timing Organization

Acknowledgments

This book project could not have been completed without the help, encouragement, and cooperation I received from my teachers and friends. I am grateful for the circle of my advisers: Karen Chapple, Steve Cohen, David Dowall, Steven Vogel, and most invaluably, AnnaLee Saxenian.

I was privileged to meet a cluster of great scholars and friends at Berkeley, and am grateful for the relationship and friendship since then: Manuel Castells, Mike Teitz, Robert Cole, T.J. Pempel, Martin Kenney, Susan Helper, David Mowery, Peter Evans, and Youtien Hsing. Friends include Makiko Takekuro, Keiko Hjersman, Victor Polanco, Alberto DiMinin, Gary Fields, Lee Chung-Kai, Xiaohong Quan, Liang-Chih Chen, Wang Jenn-Hwan, Qing, Bradley Flamm, Neil Hrushowy, Pitch Pongsawat, Vikki Chanse, Kenji Kushida, and Yumiko Mikanagi.

Shuzo Katsumoto opened the door to my research by coordinating interviews with his company. I enjoyed the informal conversation with Eishi Endo that first hinted to me about the existence of an unglobalized dimension of R&D. Yukihisa Hayakawa and Jennifer Y. Ishiguro provided insights and coordination with their company. Aki Shoji was a brilliant professional who tested and supported my earlier ideas. I appreciate all interviewees for their cooperation despite their being so busy.

I am fortunate to have had an even greater circle of intellectuals after I left Berkeley. Michael Storper's seminar became an important asset to me. I appreciated comments about a chapter from David Rigby and suggestions about publication from Alan Scott and Ann Markusen. Beth Chapple and Toddie Downs patiently reviewed my manuscript. Katy Crossan and Lianne Sherlock always responded promptly and coordinated professionally. Matt Eisler, Mikael Johansson, Jennifer Rogers, and Gwen D'Arcangelis were wonderful post-doc colleagues. Rich Appelbaum, Barbara Harthorn, and Patrick McCray were excellent supervisors.

Lastly and most importantly, I have to thank my family. My parents supported me throughout the graduate school, as well as the post-doctoral period. My wife, Michelle Johnson, was always the first and last person that I could consult with my ideas. It was the intellectual dinner discussion I could have with her, the morning smile she gave me, and afternoon tea conversation we shared in our study room that inspired and improved this manuscript. Those "intellectual" and "private" times seem a distant memory, as we now have two little children, Kai and Elyse, but they give me the inspiration to write, to research, and to live. Thank you.

August 2012
Lawrence, Kansas
Y.M.

Preface

The research in this book started as a personal inquiry. I was a doctoral student at the University of California, Berkeley, and was eager to explore the grand concept of globalization. Given my interests in the high-tech sector and research and development (R&D) activities, I wanted to see where and how firms globalize their R&D operations.

Berkeley was a global laboratory for me. With its world renown, its strengths in engineering and the natural sciences, and its prime location in Silicon Valley, the university hosted engineers and managers from companies around the world. With my Japanese background, the doors were open to some of the most globally operating firms. I was fortunate to meet people from Sony, Hitachi, Olympus, and NTT, to name a few, and I started to ask questions. How does your firm globalize your R&D activities? What does your R&D center in Silicon Valley do? What products came out of the overseas R&D labs? How do you coordinate the process?

It did not take long before I discovered that my questions were somewhat at odds with their practices. I was never told that a product or technology that represented the core business of a company had been globally produced. Overseas R&D centers were operating, but rather vaguely and with peripheral functions. It was not as though those Japanese firms lacked a global R&D strategy. Instead, it was simply better to describe their R&D operation as not being global.

I decided to narrow my question to the following. Who, within each company, had developed recent well-known and successful products? Where were those researchers located? If there is geographic concentration, why so? Initially I asked about the Vaio laptop by Sony and a new inkjet printer by Canon. The answers were strikingly similar: This product was made by this team in Japan; that technology was developed by another team in Japan. These inquiries confirmed to me that virtually every product was developed in Japan. No major product or core technology came out of foreign (outside Japan) R&D labs. Moreover, I did not hear of a single intra-firm international collaboration that successfully produced a tangible output for end users. In short, Japanese R&D activities are locally conducted.

My inquiry showed me that I had to approach the term *globalization* carefully and critically. Those Japanese companies were global in the sense that they sold products worldwide and operated manufacturing facilities in a number of countries. However, their R&D activities were highly geographically concentrated not only within Japan, but also within the firm's core location. There was a strange mixture of globalization and localization, termed *glocalization* by Akio Morita, former chairman of Sony. This book will explore how these global companies operate their product development activities locally.

Chapter 1

Introduction

It is the era of globalization. Multinational corporations (MNCs) conduct production, distribution, and sales of goods and services anywhere in the world. These are not a small portion of economic output; MNCs account for 20–30 per cent of total world output and 66–70 per cent of world trade (UNCTAD 1997, 1998). Scholars argue that global production networks facilitate product design and production technology (Ernst 1994, 5–6; Castells 2000, 207). It is also the age of information technology. People can exchange information instantly anywhere on the planet and collect any kind of data via the Internet.

However, a closer investigation into how Toyota, Sony, and Canon, three successful global players, conduct their R&D activities reveals that the overwhelming proportion of R&D takes place within each firm's core region. Moreover, the scope of this core region is substantially smaller than national, or all of Japan, but rather matches the metropolitan scale: the Toyota-Nagoya region for Toyota, and the southern Tokyo region for Sony and Canon. All the major technologies and products come from R&D divisions at the core location. This poses a puzzle because R&D activities are a critical aspect for innovation and profit making for firms, yet they do not include knowledge from around the world. Furthermore, it is against the conventional wisdom and the so-called "death of distance" theory, which argues that location is no longer important (O'Brien 1992; Cairncross 1997, 2001).

We have to approach the term *globalization* carefully and critically. These Japanese companies were global in the sense that they sold products worldwide and operated manufacturing facilities in a number of countries. Yet, they conducted R&D activities in highly concentrated areas. There was a strange mixture of globalization and localization within these global companies.

This odd combination of localization and globalization exhibits two disparities: the theory versus the practice of globalization, as well as that between what companies say and what they do when it comes to global development. First, there is a gap between the academic concept and the companies' practice in the globalization of R&D. Past research in international business management differentiated between multinational and global operations (Nohria and Goshal 1997). Multinational (or multidomestic) operation of R&D means that a firm has R&D centers in several countries, but they specialize in different technological fields or projects. In contrast, the global operation of R&D means that such R&D centers simultaneously coordinate and conduct the same project or develop the same technology together. While it is easy to theorize about a global R&D operation, I found none in practice. Real companies had only multidomestic R&D operations.

The second gap is present within each firm. These firms try to promote themselves as global companies with global R&D networks. Sony (2004b) explicitly calls for "a global synergy in R&D," and its former non-Japanese president aims for a "global cross-company team" (Stringer, cited in Sony 2006, 3). Likewise, Canon's "R&D centers around the world try to develop creative products and solutions for Canon as a whole" (Canon 2006f), and Toyota explicitly promotes technology and product development at the global level in its Global Vision 2010 (2002) and Diversity Management (2003, 19) publications. Yet, again, in their practice, R&D operations are not global, but multidomestic at best.

Why do those so-called "global" firms operate R&D activities locally? Why is that local operation geographically concentrated at each firm's core location? What is special about the nature of R&D compared to production, distribution, and sales? What exactly are the advantages of proximity for R&D activities? What is the role of information technology in this process of R&D, as it is often claimed that IT connects and disseminates ideas globally? This book explores the answers to these questions.

In Search of Answers

Past studies in business and innovation studies, economic geography, and urban planning have investigated research questions similar to that proposed here about the geographic concentration of industrial activities. Chapter 2 describes three scholarly traditions, how each school of thought answered the question, and their limitations.

The first school of thought was derived from Marshallian agglomeration theory and identified thick institutional connections within a region (Castells 1989; Lundvall and Johnson 1994; Cooke and Morgan 1998; Porter 2000). The second emphasized the importance of trust in social networks and face-to-face interactions (Sabel 1993; Saxenian 1994; Storper and Venables 2004). The third applied the theory of tacit knowledge (Polanyi 1966; Nonaka and Takeuchi 1995), and argued that some knowledge did not travel over distance and had to be shared by collocated people (Gertler 2003; Zook 2004).

Despite this collection of debates, the major limitation of past studies is that they observed innovation at the aggregated, regional level. Hence, scholars could treat innovation only holistically and discussed it along with macroeconomic proxies, such as the growth of the high-tech sector in Silicon Valley (Saxenian 1994; Kenney 2000), the number of patents produced in a given region (Jaffe and Trajtenberg 2002), or the presence of creative professionals (Florida 2004). All of these phenomena had some correlations with innovations, but they were not innovations per se. As a result, these studies were unable to investigate the inside of innovation and technology, particularly the engineering process of generating each innovation and its connection to a specific location. Innovations have remained a black box, as Nathan Rosenberg (1982; 1994) termed it decades ago.

In contrast, this book attempts to open this black box and argues that the specificity of innovations matters and that it is essential to examine the mechanism that creates each innovation. Contemporary innovations often come from engineering and technical matters. Without understanding the specificity and the technical aspect, we may be missing the fundamental nature of the process of creating innovation and the connection to a particular location. The microdynamic analysis of this book shows why the specificity and complexity among different innovations requires collocated development. Applying the historical corporate case study method pioneered by Hounshell and Smith (1988), we will investigate how each project started, who was involved, where the participants were located, what kinds of technological difficulties they faced, and how they struggled and solved problems. Then we will evaluate why each of those processes required localization or collocation.

Studying Japanese multinational firms can provide an advantage in investigating technology making. Sony, Canon, and Toyota are excellent case companies because each is a major global player in its respective industrial sector: electronics-semiconductors, optics, and automobiles. These are successful firms in a world market, with overseas R&D units. In other words, these firms have strong intentions to develop and sell their products at the global level. Moreover, the business and the core competency of each firm are rooted in the innovation- and engineering-intensive technological fields.

At the same time, it is important to consider whether these Japanese firms present a unique pattern compared to multinationals from other countries. Indeed, some studies in international political economy (Ernst 2000; Katzenstein 2003; Keller and Samuels 2003) and international business (Cantwell 1995; Lam 2003; Urata 1999; Westney 1999) have argued that Japanese firms moved their R&D only selectively and marginally overseas. Chapter 2 additionally introduces these two streams of literature and their reasons for claiming Japanese firms are an anomaly. However, they lack a systematic framework to compare multinationals among different countries of origin. More comparative works (Dunning 1992; Doremus et al. 1998) revealed that both U.S. and Japanese firms conducted the overwhelming proportion of their R&D, approximately 90 per cent, in their home country. Additionally, recent, in-depth industry studies have supported the home-country bias of R&D activities by U.S. firms in various sectors: semiconductors (McKendrick, Doner, and Haggard 2000), automobiles (Studer-Noguez 2002; Sturgeon and Florida 2004), telecommunications (Cohen et al. 2009), and biotechnology and software (Macher and Mowery 2008). Therefore, the pattern of geographic concentration of R&D applies to non-Japanese firms as well, and findings from this book will have important implications not only for Japanese, but also for non-Japanese firms with regard to the global organization of R&D and the engineering aspect of innovation making.

Plan of the Book

Chapters 3–5 describe how the three firms developed their successful products. With a detailed, microlevel analysis, each chapter relates stories of the engineering and technical aspects of product development.

Sony started to develop its Vaio 505 in November 1996. Initially, one designer was located at its headquarters, Shinagawa, in central Tokyo, and five engineers worked at a technical center in Fujisawa, which is 45 km (28 miles) southwest of the headquarters. At this early stage of concept and component development, the designer from the headquarters traveled to Fujisawa for every meeting. One meeting room was dedicated to this project team, and there were almost always some members analyzing the concept and paper models in the meeting room. Gradually, more engineers joined the project in Fujisawa, increasing to eight in February and to 15 in September 1997. Essentially, Sony's expertise in electronics, mechanical and software engineering, materials science, and design contributed to the miniaturization of a laptop computer. When the process shifted to mass production, most meetings moved to their headquarters. Production took place in Nagano, 250 km (156 miles) northwest of Tokyo, and the product came onto the market in November 1997. R&D labs at San Jose and Brussels participated in this product development, but their roles were restricted to local adaptation of the product, including the translation of the brochure and labels.

Canon started to explore alternative inkjet printing technology as early as 1977. This required basic research, and the 20 members of the project initially used the lab facilities of Tokyo Metropolitan University at Meguro, in central Tokyo. Over time, this team moved through three locations: the company's central research lab at Atsugi (1979–81), the technical center at Hiratsuka (1981–89), and eventually to the headquarters in Shimomaruko (1989–present). All these places were located within 50 km (31 miles) of the company headquarters. The project members increased to over 200 by 1987 and 300 by 1989, when in fact all these members were located at the same place. Canon had to mobilize all the knowledge it possessed in electrical and precision machinery, physics, and simulation engineering. Only after production began did the locations used spread to non-Tokyo metropolitan areas. There was no involvement of overseas R&D labs in the U.S., Europe, or Australia.

Concern expressed by Toyota's Chairman led to the formation of a research group to explore concepts for cars of the 21st century in September 1993. Interestingly, the concept drawn up by this group in December had nothing to do with the hybrid engine, but focused on three broad themes: small size, high fuel efficiency, yet spaciousness. Toyota worked further on specifying the concept, but it was not until November 1994 that three executive figures decided to apply hybrid technology to the car for the next century. Then, Toyota quickly scaled up product development teams, almost all at its development center in Toyota City, including power-train, chassis, body engineering, interior engineering, and vehicle evaluation groups. Only the group developing the engine system integration was located outside the city in Higashi Fuji, 240 km (150 miles) east of Toyota City. However, the development group realized that the two-hour journey was a physical burden and moved the engine system integration team to Toyota City by January 1996. Additionally, together with its coinvestor and long-time ally, Matsushita Electronics, Toyota established a new firm to create a new type of battery in Kosai, 75 km (47 miles) southeast of Toyota, in December 1996. The Prius was indeed a hybrid product of Toyota U.S.A. and Japan. A team at Calty Design Research in Los Angeles modeled the exterior design throughout 1996. However, this design was almost independently conducted. For automobiles, the exterior design must appeal to consumers, but that does not affect the development of other components once the basic body size is fixed. Production took place in four factories within Toyota City. The executives pushed the introduction of this car before the new century, finally bringing the hybrid car to market in December 1997.

In the last chapter, "Innovation and Geography," I will draw conclusions for the management of innovation at the global and local levels. Based on the findings from the case chapters, I will demonstrate that three engineering and organizational aspects of innovation making necessitated the collocation of each firm's development resources and personnel. The three aspects are: the complexity of modern consumer products, the interdisciplinary dimension of technological advancement, and the need for prototyping and testing. These microdynamic aspects were the hidden dimension of innovation making that previous theories of agglomeration, social networks, and tacit knowledge have overlooked. At the same time, this microdynamic theory will not seek to refute other theories, but to complement them. This analysis can deepen the understanding of why social networks and tacit knowledge stick to places. There will be further discussion in relation to the theory of innovation, agglomeration, and firms.

This book will challenge the conventional understanding of globalization, whether R&D has been globalized, and provide a nuanced conclusion. Multinationals do have overseas R&D centers. However, the overwhelming preponderance of R&D was conducted at the core location of each multinational, and overseas R&D centers played only a peripheral role. Limited and peripheral inputs came from overseas R&D centers in Sony's and Toyota's cases which is a substantially different picture from the conventional understanding of the globalization of R&D and the claims by multinationals. Multinationals may have R&D centers with a global stretch when simply looking at location, but organize them in a hierarchical and multi-layered structure.

The implications from these findings are substantial. While access to world-class talent and emerging technology clusters led to a rush to build R&D centers by multinationals in China and India (for example, Reddy 2000; Friedman 2005; Boutellier 2008), and these countries indeed achieved remarkable growth of high-tech industries in the last decade, this book will suggest that the operation of global R&D and the synergistic integration of knowledge, innovations, and technologies is not a simple task. Outsourcing of R&D may take place, but only if a component or functionality is specifically defined and separated.

Information technology facilitated the product development process, but did not substitute for collocation. Firms applied information technology to complement the advantages of proximity, in other words, to make the innovation process more efficient and effective, but hardly had any interest in replacing the advantages of collocation.

This book concludes that innovation is not only a product, but also a process of trial and error with numerous open-ended possibilities. Therefore, firms have to plan, ironically, for the unplanned consequences of product development. Because of this uncertain nature and the need to change and adjust constantly, R&D activities are fundamentally different from other corporate functions, such as production, distribution, and sales, which the multinational firms can locate overseas and coordinate from afar. We will see how each case unwraps this microdynamic process of innovation making.

Chapter 2
In Search of Answers to Being Global and Local

This chapter has four objectives. First, to examine the concept of globalization which is a term that has been used and defined in so many ways that it can mean almost anything. Since the primary focus of this book is on globalization and its contrasting phenomenon, localization, we have to be careful when discussing the concept, or we will not reach any agreement. By assessing several definitions, the first section will clarify this book's scope of globalization and explain why that focus matters when analyzing innovation and geography.

The second objective is to investigate potential answers to the question: why do global firms operate R&D locally? There are three schools of thought that could provide potential answers: agglomeration, social networks, and tacit knowledge. Each school's potential for explanation will be assessed, followed by consideration of what is missing in their approaches, and how this book will complete the analysis.

Third, we will assess the advantages and disadvantages of studying Japanese firms, as well as the implications for U.S.-based multinationals. Fourth, the methodology of this study will be briefly explained.

2.1 Which Aspect of Globalization Are We Talking About? What is Globalization?

Globalization emerged as a buzz word in the 1990s (Cooper 2004, 152; Evans 2002, 1; Keohane and Nye 2000, 104) and is a concept frequently used by so many: politicians, media reporters, social workers, street gangs, and others. Among academics, the concept has a whole range of connotations, from an integration of commodity, labor, and capital markets in economics (Bhagwati 2004; Bordo, Taylor, and Williamson 2003), the diminishing role of the nation-state in political science (Cable 1995; Ohmae 1995; Schmidt 1995; Strange 1995), the expansion of capitalist ideology in sociology (Ferguson 1992; Bourdieu and Wacquant 1999), or the compression of time and space (Cairncross 1997, 2001; Harvey 1989; O'Brien 1992). Anthony Giddens (2000) and Mauro Guillen (2001) attempted to integrate various theories from the social sciences and defined globalization as the convergence of cultural, political, and economic aspects of life.

This chapter does not discuss each of these concepts or aim to come up with a single grand theory of globalization. Simply put, no such thing exists because globalization has so many faces. However, it does not mean that we can talk about globalization however we want. It is more important to clarify which dimension of globalization we will discuss. As Bauman remarked, "vogue words share a similar fate: the more experiences they pretend to make transparent, the more they themselves become opaque" (1998, 1). It is important to agree on which aspect of globalization we will discuss, otherwise, we can have neither a single destination nor even a common map to explore with.

This is a book about companies and their innovations. Hence, we will focus on the economic aspect of globalization. But, even within the economic aspect, the concept of globalization varies substantially. This section introduces several concepts concerning economic globalization. The bottom line that I would like to reach is the following: When we talk about globalization, we are not concerned about the quantitative increase in international economic activity. In other words, we are not interested in how many factories and R&D centers Toyota or other companies established in foreign countries, since this hardly captures the dynamics of contemporary economic globalization. Instead, what we will focus on is the qualitative linkage within each multinational corporation: how and how well the R&D centers in multiple countries are coordinated.

2.1.1 Trade

The globalization of economic activity is often thought to have appeared after World War II (Hirst and Thompson 2003, 335), and its pace has increasingly accelerated since the 1980s. In fact, before the 1990s, the term globalization "was hardly used, either in the academic literature or in everyday language" (Giddens 2000, 7). With the end of the Cold War, globalization seemed to have come from nowhere to be almost everywhere in the 1990s. That globalization as a phenomenon is recent is a common misconception. This section will explain the error with three indicators: trade, multinational corporations, and immigration.

The Organization for Economic Cooperation and Development (OECD) is an intergovernmental think tank specializing in research into the policies and economies of the industrialized countries. Their statistically rich reports complement this conventional view of globalization. Assessing trade in goods and services, foreign direct investment, and multinational corporations, OECD concluded that the intensity and multiplicity of economic linkages accelerated in the past two decades (2005a, 18; 2005b, 16). The following figure (Figure 2.1) summarizes the typical perception of the trend.

Between 1995 and 2003, all nations experienced an increase in trade as a percentage of GDP. The degree of trade penetration varied by country: from 135 per cent for Netherlands to 25 per cent for Japan in 2005, but the increase was steady and across the board. There seemed to be a clear pattern of increasing economic globalization.

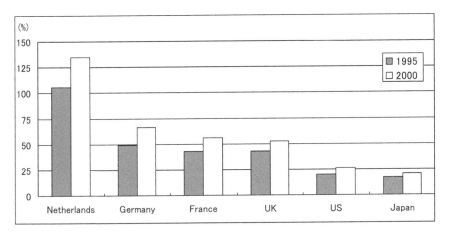

Figure 2.1   Trade as a percentage of GDP[1]

Source: OECD (2009).

However, this analysis of trade and the conclusion about globalization was called into question by several scholars (Hirst and Thompson 1999; Sachs and Warner 1995; Zevin 1992) as they analyzed the same statistics over a longer period of time. Compared to the beginning of 20th century the level of trade activity in 1995 was not unprecedented, but rather at a level comparable to 1913 (see Table 2.1 below). France, Germany, and the U.S. seemed to have a higher level of trade in 2000, while it was essentially the same in the Netherlands and the U.K., and the rate was lower in Japan.

Looking at the big picture, this economic integration by trade took place between approximately 1850 and 1914, but the pattern reversed itself between 1914 and 1945. It is only since 1950 that there has been consistent expansion of trade (Keohane and Nye 2000, 107; Hirst and Thompson 2003, 336). It is also true that the level of trade has increased since the 1990s, and almost all industrialized nations, except Japan, experienced the highest level by 2000. However, it is difficult to argue that we have entered an era with a fundamentally different level from the past.

---

1   As the scale of economy expands over years, the total volume of trade increases naturally. We have to measure the trade by the size of economic activity, i.e. GDP. Here, trade is measured as a sum of the imports and exports of goods and services.

Table 2.1   Trade as a percentage of GDP from 1913 to 2005

	1913	1950	1973	2000
France	35.4	21.2	29.0	56.2
Germany	35.1	20.1	35.2	66.4
Japan	31.4	16.9	18.3	20.5
Netherlands	103.6	70.2	80.1	134.6
U.K.	44.7	36.0	39.3	52.5
U.S.	11.2	7.0	10.5	26.2

Source: Calculated from Maddison (1987, 695) for 1913 to 1973; OECD (2009) for 2000.

2.1.2 Multinational Corporations

The second misconception about globalization is about the role played by multinational corporations.[2] Often, it is considered that these powerful organizations emerged over the last few decades, played a considerably large role in world economy, and accelerated the pace of globalization. Undoubtedly, multinationals are among the most powerful organizations in world economy in terms of the availability and mobility of capital and human resources (Castells 1996, 119, 126). They account for 66–70 per cent of world trade (UNCTAD 1997; Held et al. 1999), and one-third of cross-border trade in non-agricultural goods and services takes place within multinational corporations (Bordo, Eichengreen, and Irwin 1999, 12). This figure increases to 50–60 per cent of capital and technology flows in the transactions of the world economy (Dunning 2000, 14).

However, the history of multinationals goes back centuries—they are not a creation of the modern era. In the Middle Ages, Italian bankers such as Bardi and Peruzzi operated in England (Wilson 1976, 265). The British and Dutch East India Companies were founded in 1601 and 1602, respectively, and traded goods from the other side of the planet (Wilkins 1991). In addition to these service and trading activities, multinational manufacturing activities started soon after the Napoleonic Wars in the 1810s, when British firms operated textile factories and later constructed railroads in France (White 1933). These activities expanded to a wider variety of industries as Swiss, German, Swedish, British, and Belgian companies produced goods such as baby food, soap, margarine, aspirin, glass, alkali, and ball bearings (Franko 1976; Wilkins 1977).

---

2   Some scholars use the term "transnational" corporations. Throughout this book, I will use *multinational* and follow the simple definition by Wilkins: an organization that "makes foreign investments and establishes stakes abroad that are under the control of the parent firm" (1977, 577).

Unfortunately, unlike the data we have for trade, there is no available data systematically measuring the scale of multinationals over the total economic volume for a century. The best statistics[3] are provided by Zeile (1997), who reported that the ratio of intrafirm trade to total trade was stable between 1977 and the late 1990s. There is no evidence to suggest that multinationals play a larger role in world economy today. In general, historians agree that the modern form of multinationals appeared after the mid 19th century and was well established by the First World War. However, this international business activity slowed down during the depressed 1930s and war-torn 1940s, while expansion started, with periodic fluctuation, after the 1950s (Dunning 1992; Hirst and Thompson 2003, 336; Wilson 1976). Essentially, this is the same picture as the level of trade. At the minimum, we can conclude that the emergence of multinational firms or the increased number of them does not uniquely characterize today's economic globalization.

2.1.3 Immigration

The third misconception about globalization has to do with immigration. It is often thought that the migration of people has recently reached the highest level in human history. For instance, one in eleven people in the United States today is foreign-born, and more than 50 million are recent immigrants and their children (Hirschman, Kasinitz, and DeWing 1999, 1). However, this is another misconception because both the number and percentage of immigrants were greater in previous years.

In the United States, over 1.2 million people immigrated annually during the early 1900s, declining dramatically to 50,000 in the 1930s and early 1940s. This figure started to increase in the post war period; however, it was still less than 0.7 million in 2000, approximately the same level as the 1880s (Carter and Sutch

---

3   Another common method to measure the activity level of multinational corporations is foreign investment. But this data has two potential problems. First, foreign investment takes two forms: portfolio investment, involving the acquisition of foreign securities without any control over the management of the company, and foreign direct investment (FDI), involving management control. There is no consistent definition among countries as to how much the minimum equity stake is. For example, in the U.S. and France, 10 per cent of equity is considered as FDI, while it is 20 per cent in Germany and Britain (for more details, see Jones 2005, 7). Second, there is no comparable data available from the beginning of the 20th century. Some scholars estimate that most of the foreign investment back then was portfolio (Dunning 1983, 1992; Bordo, Eichengreen, and Irwin 1999), but data from the then largest economy, Great Britain, is not available (Platt 1980; Feinstein 1990). Moreover, UNCTAD (1994, 1999) estimated that the ratio of FDI to total world output did not vary substantially between 1913 (9.0%) and 1997 (11.8%). Therefore, there is no convincing evidence to claim that the multinationals play a larger role today.

1999; Census 2000; Serow et al. 1990).[4] After 9/11 in 2001, the regulations on immigration were tightened substantially; the number of immigrants continues to decline. Consequently, the ratio of immigrants to total population, and the immigrants' contribution to total population growth, were both greater in the 1850s, 1880s–1890s, and 1900–1910s (see Figure 2.2).

The greatest era of intercontinental mass migration started in the early 19th century and continued for about five decades. More than 60 million people left

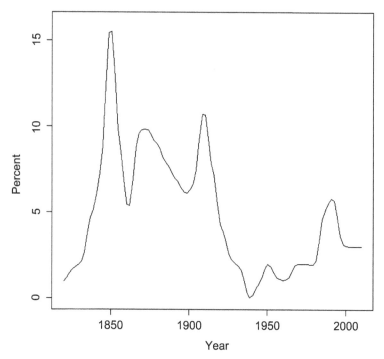

Figure 2.2   Net immigration contribution to population growth => Ratio of legal immigrants to population

Source: McClleland and Zeckhauser (2004) for 1860; Kuznets and Rubin (1954) for 1871–1939; Census Bureau (2000) for 1940–2000; modified from Carter and Sutch (1999, 321).

---

4   By definition, there are no official statistics about illegal immigrants, so it is always difficult to estimate their number. Estimates include 8 million by the Census Bureau (Costanzo et al. 2001), 11.3 million by the Center for Immigration Studies (Camarota and Krikorian 2008), and 11.7 million by the Pew Hispanic Center (Passel and Kohn 2008). This still amounts to only 2.7% to 3.9% of the U.S. population and does not significantly affect the discussion in this section.

Europe for the Americas or Oceania, and additionally 10 million Russians, 12 million Chinese, and six million Japanese left their homelands (Segal, Chalk, and Shields 1993). This was the age of open migration. Since the 1960s, the policy environment in major countries has changed to the opposition of immigration. Only highly skilled professionals have the option to move, while the poor and low-skilled immigrants have few possibilities. The present day may be described as the era of constrained migration, ironic given the advancement of transportation technology (Chiswick and Hatton 2003, 650; Hirst and Thompson 2003, 339).

2.1.4 Finding What's New in Today's Globalization

We have examined three indicators most commonly associated with globalization: trade, multinational corporations, and immigration. All three cases present a similar picture: Modern globalization started in the early to mid 19th century and accelerated its pace toward the early 20th century. It was disrupted during the two world wars, but started to recover slowly in the post war years. People have often debated the trend of globalization since the 1980s, when a rapid acceleration began, but that analysis was myopic. Compared to the world economy in the 1910s, we are not living in a fundamentally different era of globalization. This assessment contradicts the conventional understanding of globalization. Yet, it is true at least if we examine the quantitative level of economic activity.

Then, is there anything new about contemporary globalization? Are we living in the same kind of world as a century ago? What roles do information technology and advanced transportation play, since the Internet and jet airplanes were not available back then?

It is easy to conclude that there is nothing new under the sun, especially if we focus on the quantitative aspect of economic activity. There is no absolute discontinuity in human history, and a fundamental change is not an easy assessment to make. Even if today's trade activity is 10 or 30 per cent higher than a century ago, is it different enough to declare a new era? There is no absolute answer to this question.

A framework presented by Keohane and Nye (2000, 108) and Held et al. (1999) may be useful in differentiating contemporary globalization. They argued that the major difference came out of a qualitative shift from *thin* to *thick* relations. A thick connection required both *intensive* and *extensive* networks. For instance, since the 19th century, the time for trans-Atlantic communication has shortened dramatically, first, by steamships (a few weeks), then by telegrams (a few minutes with decoding), next by telephone (almost instant), and finally by the Internet (about the same). These changes in velocity increased the intensity of connections. However, it also increased the extensity of connection. What matters here is not the point at which the two continents were connected by trans-Atlantic cable, which took place in 1866. At that time, it was too expensive for ordinary citizens to send telegrams, and only corporations and the rich used the telegraph. Moreover, they did so only for messages regarded as urgent. In contrast, by the 21st century, international telephone calls cost only cents, and emails and other Internet-based communication are virtually free. This has fundamentally changed the extensity of networks because hundreds of millions of people now communicate over thousands of miles. It has further augmented the intensity of networks because people can exchange information from the most strategic information about corporations to personal affairs on a daily basis, regardless of which is more important. Therefore, what we perceive as today's globalization is the thick intensity and extensity of relations.

This framework critically shapes how we can look at multinational corporations. The quantitative aspect, such as the presence or the simple number of such corporations, does not capture what contemporary globalization means. In contrast, we will focus on the qualitatively thick linkage between them. Thus, in this book, we will look at *how multinationals are coordinated in real time on the intercontinental scale.* For instance, this means that a change in demand in North America can affect procurement in Asia on the same day.

We further apply this framework specifically to the research and development (R&D) activities of multinationals. Multinationals can have R&D centers in multiple countries, what is conventionally called their "global R&D network." However, with our framework, this is not a sufficient condition to be labeled as contemporary globalization. Each R&D center must have qualitatively thick coordination. This must be more than sharing the same corporate values or for the headquarters to supervise the overseas R&D directors. R&D centers in different locations must work on the development of the same or a complementary product or technology with real-time coordination. If each R&D center pursues different projects on different technologies, the operation is best described as multidomestic, not global.[5]

Using this definition of globalization, Sony, Toyota, and Canon are not globally operating their R&D centers. Their major operations are based in their core, home regions: the southern Tokyo metropolitan area for Sony and Canon, and the Nagoya-Toyota area for Toyota. Overseas R&D centers were working on small and different development projects, or supported their main R&D center for somewhat peripheral functions. As a result, all their major products and technologies came out of their home regions. In this book, we will investigate the reasons for this geographic concentration of product development activities.

2.2 How Could Other Theories Answer the Questions?

This section will review potential answers to the central research question of the book: Why do those so-called "global" firms operate R&D activities locally? What exactly are the advantages of proximity for R&D activities? Previous research

---

5   Nohria and Ghoshal (1997) used the term "transnational," which is essentially the same as what I mean by global. To avoid excess jargon, I stay with global throughout the book.

has identified some similar questions, and can be categorized into three different schools of thought.

2.2.1 Agglomeration Literature

The first school, agglomeration, has the longest tradition and investigated why certain industrial activities are geographically concentrated. Analyzing the high concentration of the textile industry in Manchester and the cutlery industry in Sheffield, among others, Alfred Marshall found that firms benefited from colocation by sharing the pool of highly skilled labour, customers, specialized suppliers, and ancillary services (Marshall 1898, 346–56). The emergence of electronics and semiconductor industries in the 1970s brought a new series of studies. The creation of innovation came increasingly from broader economic entities, such as collaborations among firms, research organizations (universities and other public research institutions), and public agencies (technological transfer centers, economic development agencies) (Asheim and Gertler 2005, 293; Castells 1989; Porter 1994, 2000). This linkage between economic actors within a region was fundamental to growth and innovation. The competition and collaboration among the economic actors fostered what scholars called "learning through interacting" (Lundvall and Johnson 1994; Florida 1995; Cooke and Morgan 1998).[6] In short, the school of agglomeration asserted that innovation came from the interaction among different institutional economic players in the same region inducing the development and adaptation of specific technologies in a region (Storper 1997; Scott 1998).

2.2.2 Social Network Literature

The second school of thought, social networks, was based on the influential work by a sociologist, Mark Granovetter (1985), who revealed that economic relations were often underpinned by personal relationships instead of purely rational calculation of the market. These relationships developed based on trust, which Charles Sabel referred to as "the mutual confidence that no party to an exchange will exploit the others' vulnerability" (1993, 104). Sabel further claimed that a social and economic environment based on trust was a precondition for cooperation alongside competition and provided mutual benefit for firms and individuals. This helped to explain why some regions which internalized the promotion of trust were able to prosper better (Chinitz 1961; Brusco 1982; Piore and Sabel 1984; Putnam, Leonardi and Nanetti 1993). According to Saxenian (1994), Silicon Valley outgrew Boston, a traditional cluster of high-tech semiconductor firms, because of its unique industrial structure and institutional environment. The

---

6   Audretsch and Feldman (1996), Feldman (1994), Jaffe et al. (1993), and Zucker and Darby (1996) and Zucker et al. (1998) concluded similarly, while their main analysis was the knowledge spillover effect mediated by geography.

industrial structure of Silicon Valley was more disintegrated, with small venture start-ups, and such an institutional environment promoted a less hierarchical and more open environment for information exchange. While the first agglomeration school emphasized the formal institutional structure, such as firms and universities, the second school of thought focused on personal networks and the informal and regionally embedded culture.[7]

This second school of thought additionally stressed the importance of face-to-face interaction. People relied on social networks to assure the source and quality of information (Currid 2007). Face-to-face contacts were the most reliable mode of communication to detect emotions, willingness to cooperate, and confidence, as well as to promote the iterative process of judging and being judged (Storper 1995; Storper and Venables 2004). This face-to-face communication, by definition, required colocation.

2.2.3 Tacit Knowledge Literature

The third school of thought applied the theory of tacit knowledge. Michael Polanyi (1966) argued that knowledge had two dimensions: codified and tacit. One could write codified knowledge relatively easily, while one could not express tacit knowledge explicitly. Riding a bicycle was the classic example of the latter. A person certainly knew how to ride a bicycle, yet could not explain every detail of it.

This division of knowledge had two significant implications for innovation and geography. First, success under economic competition depended increasingly on the ability to produce new or improved products and processes, and the creation of new knowledge was the most essential source for this (Pavitt 2002). The fact that we are living in the information age does not mean that everyone is better off. When everyone has relatively easy access to codified knowledge, the creation of unique capabilities and products depends more on the production of tacit knowledge (Malmberg and Maskell 1997, 28). Moreover, Nonaka and Takeuchi (1995) argued that the process of knowledge generation required the dynamic interaction between codified and tacit knowledge, which came through the interaction within and between organizations.

Second, the division into two kinds of knowledge implied that people could disseminate codified knowledge more easily over space, in the form of newspapers, multimedia, the Internet, and academic journals, but would have more difficulty transmitting tacit knowledge. At the micro-level, tacit knowledge was shared among people of similar conventions, codes, past experiences, and daily routines. At the macro-level, tacit knowledge was generated and transmitted in the social and institutional context (Gertler 2003; Zook 2004). At both levels,

---

7 Asheim (1998) and Cooke (1998) distinguished the agglomeration school as regionally networked innovation system and the social network school as territorially embedded regional innovation systems.

tacit knowledge was inherently "sticky" to space, which limited the diffusion of innovation-making activities and forced colocation. In other words, the unexpressed knowledge bonded people together in a region and allowed higher productivity of innovation-making.

2.2.4 What is Missing?

What is missing in these schools of thought? While they all provided profound insight into the relationship between innovation and geography, they provided little understanding of the actual mechanism of how innovation was created. We know that innovation is geographically concentrated. We further know that proximity and interactions among economic agents are essential for innovation, regardless of involving social networks, tacit knowledge, and its spillover. However, the actual process of creating innovation must be deeper and more complex. Using Marshall's famous expression, the process of making is still "in the air" (1898, 350). This mechanism essentially remains "a black box" (Rosenberg 1982, 1994), and to date, the mechanisms of externalities and knowledge spillovers have not yet been explained (Feldman 2000, 389; Storper and Venables 2004, 352).

Being collocated is only a precondition. What exactly is happening in the interaction between suppliers, customers, and rival firms? Meeting with people face-to-face is undoubtedly essential, but what happens after they have developed trusted social networks? If people work in the same organization and share routines and tacit knowledge can they later be separated geographically and still create innovation? (Keep in mind that it is extremely difficult to observe this with tacit knowledge, by its definition.)

Current studies have little explanatory power to address these questions or to explain the mechanism of making innovation because they do not investigate the specificity of innovation. Innovation takes many forms: a new commodity or a qualitatively better version of existing ones, a new market, a new method of production, or a new organization (Schumpeter 1934, 66). I suspect that every one of these requires a different process. Unfortunately, since past studies of innovation and geography often analyzed innovation at the national or, at best, regional level, they could only treat it abstractly and holistically. Innovation was measured through rough proxies such as the growth of high-tech industries, expenditure on R&D, or the number of patents applied for.[8]

As a result, these studies were unable to investigate the inside of innovation and technology, particularly the technical and engineering process of generating innovation. This is a serious limitation because contemporary innovations often come from engineering and technical matters. Without understanding the specificity and the technical aspects, we may be missing the fundamental nature of the process of creating innovation and its connection to a specific location.

---

8 For a more detailed explanation of how past research measured innovation, see Carter (2007) and Ratanawaraha and Polenske (2007).

However, in this book, we will analyze the specificity of innovation and highlight the mechanisms that create it. We will look at the microdynamics of innovation and observe how firms create new products. We will analyze how engineers and managers within a multinational corporation are organized and coordinate with their overseas R&D centers. Ultimately, we will observe the actual innovations one by one, as concretely as how Sony's engineers solved a heat issue in making a computer. Using this approach, we will not treat *innovation* as a broad and holistic term, but part of a larger series of *innovations*.

I believe that analyzing the specificity of innovation will powerfully illuminate its nature and its connection to location. Only after investigating the specificity of innovation can we start to understand the complex and heterogeneous nature of innovation, such as the long process of product development, the uncertain and open possibilities of technology, and the necessity for testing and improving. More importantly, we will discover how all these features have deep geographic roots.

2.3 Why Japanese Firms? Why Not Japanese Firms?

We have to consider both the advantages and disadvantages of using Japanese firms as a case study because this limited geography may show behavior peculiar to Japanese firms. In this case, we may apply the findings from this book only to Japanese firms. This section will introduce past studies that argued the unique behavior of Japanese firms and the limited geography of innovation. However, these studies overlooked one critical factor: U.S. multinationals also have limited geographic scale in their R&D activities. Therefore, it is not anomalous to Japanese firms, and the findings from this study will have implications for U.S. firms.

2.3.1 The Anomaly of Japanese Firms' R&D

Two groups of studies have suggested that Japanese firms moved their R&D only selectively and marginally overseas: those in political economy (Ernst 2000; Ernst and Ravenhill 2000; Katzenstein 2003; Keller and Samuels 2003) and those concerning multinational businesses (Sakakibara and Westney 1992; Cantwell 1995; Urata 1999; Westney 1999; Lam 2003). These studies essentially argued that some peculiarity of Japaneseness, whether it was nationalism or the business culture, determined this limited pattern of overseas R&D activities. Additionally, they implicitly or explicitly argued that non-Japanese firms, especially U.S. firms, conducted international R&D on a larger scale and more effectively, i.e., that they have more globalized R&D activities.

In studies of political economy, Katzenstein argued that Japanese firms and institutions were "attuned to technological developments which they followed closely, imported freely, improved greatly," yet organized internationally with carefully monitored conditions, and "[t]ypically, both government and corporations make special efforts to counteract adverse shifts in technology" (Katzenstein

2003, 220, 229). Samuels and Keller added that Japanese firms shared information in the development of technologies with conational firms and that they tended to emphasize "autonomy over dependence on foreign technology, the diffusion of knowledge among national users, and the nurturance of domestic scientific and technological capabilities" (2003, 10). This was based on Samuel's earlier work on *technonationalism* in Japan in which "the belief that technology is a fundamental element in national security, that it must be indigenized, diffused and nurtured in order to make a country rich and strong" (1994, x). In prolific works about the international production network, Ernst and his colleagues argued that the *keiretsu* system induced Japanese firms to rely on familiar methods and linkages and contained the R&D activities within Japan (Ernst 1997a, 1997b, 2000, 2002; Ernst and Guerrieri 1998; Ernst and Ravenhill 2000).

A collection of studies in international business came to the same conclusion (Sakakibara and Westney 1992; Cantwell 1995; Gassmann and von Zedtwitz 1999; Gerybadze and Reger 1999; Westney 1999; Lam 2003). Urata found that the lifetime employment system among Japanese firms led to intrafirm job rotation and a heavy emphasis on on-the-job training (1999, 158). This functioned well in the home country with its large R&D divisions, but did not accumulate any capacity in overseas R&D centers, which employed only a small number of researchers, all of whom returned to Japan after a few years.

Lam (2003) similarly analyzed the labor system and argued the problem of matching institutions between the home country (Japan) and host countries (overseas). Because of the job rotation system, Japanese employees had a strong internal focus (i.e., from the perspective of the R&D headquarters in Japan) and have lost an important knowledge source they could have acquired through linkages with local labor markets and personnel exchanges abroad. Thus, this limited their capacity to develop local innovation networks abroad, and overseas units functioned mainly as technology listening posts or highly specialized units within each firm (697–9).

2.3.2 Comparing with U.S. Multinationals

Despite the findings of these studies in political economy and international business, a more systematic comparison with U.S. multinationals reveals a completely different picture of R&D and geography. In short, R&D activities are not globalized by both Japanese and U.S. firms.

First, studies of foreign direct investment show that the marginal operation of overseas R&D applies equally to U.S. multinationals. Tracing R&D expenditure levels, Dunning (1992) revealed that only 9 per cent of R&D activities of U.S. firms were undertaken outside the United States in 1989. This was only a modest increase from 6 per cent in 1966. In contrast, Japanese firms spent 5 per cent of their budget on overseas R&D. This strikingly small figure of overseas R&D activity by U.S. firms was also found in later studies by Dalton, Serapio, and

Yoshida (1999, 8), reporting 11 per cent, and Doremus et al. (1998, 86), reporting 12 per cent.

These R&D expenditure figures may be rough and only one indicator in the globalization of R&D. Nonetheless, similar findings appeared in more in-depth industrial studies emerging in the 2000s. McKendrick and others conducted a serious, in-depth study of the hard drive industry and observed a geographical separation between development and production: "It is one thing to develop a new product, quite another to manufacture it at high volumes" (2000, 95). Thus, although U.S. semiconductor firms internationalized their production to Southeast Asia, technological development remained in the United States.

In its effort to globalize, Ford had the reputation of being the most successfully transnational among the U.S. Big Three automobile makers. However, it "continued to locate the bulk of its production and R&D operations in the U.S." (Studer-Noguez 2002, 219–20). Furthermore, the Big Three have reconsolidated their R&D functions. Detroit has experienced an industry-wide reconcentration of high-paying research, design and engineering jobs since the late 1990s (Sturgeon and Florida 2004, 76).

Cohen and others (2009) analyzed patent data and observed that all the leading telecommunication firms from the U.S. and Nordic countries continued to file the most strategic patents in their home countries. Macher and Mowery (2008) conducted a major multi-industry study commissioned by the National Academy of Sciences. The results were similar in the software, semiconductor, pharmaceutical, and biotechnology sectors: U.S. firms maintained their major R&D activities at home.

Thus, a completely different picture emerges, and we should confirm two basic points: First, the overwhelming preponderance (as high as 88 or 91 per cent) of R&D was conducted in the home country even with supposedly more globalized U.S. firms. Second, of the small residue of R&D, i.e. 9 per cent or 12 per cent, Japanese firms may conduct overseas R&D activities even less, such as 5 per cent. Therefore, the limited pattern of overseas R&D also applies to non-Japanese firms and moreover seems to be an argument for only marginal differences between nations.

I will limit the scope of this book and do not intend to generalize the findings yet. At the same time, we should keep the gate open. There is no theoretical and empirical basis on which to suggest that the tight geographic circle of research activities applies only to Japanese firms. The aim of this book is to partially open the black box by using a case study of Japanese firms. While the full answer about U.S. multinationals requires further research with a comparable microdynamic approach, we have a reasonable assurance that our findings *could* apply to U.S. firms based on the similar empirical evidence of their geographic concentration of R&D activities.

2.4 The Plan of Research

The next three chapters will investigate three product development activities by three companies: the Vaio laptop by Sony, the Prius hybrid car by Toyota, and the Bubble Jet printer by Canon. These are three excellent case companies because they are each the leader in an innovation-intensive sector: electronics-semiconductor, automobile, and optics, respectively. They are successful firms in a world market, with overseas R&D units. In other words, these firms have strong intentions to sell and develop their products at the global level. Nonetheless, demonstrating the fact that these firms organize development activities highly at the local level and analyzing why can highlight the reasons for such localness.

I chose these products for three major reasons. First, each product had fundamentally new features, which would require a set of new innovations, and had a significant impact on the market. Second, each product was a high priority project for the company, which would potentially mobilize various resources within a multinational firm across divisions and possibly across overseas R&D units. Third, each product was sold worldwide in high volume.

Sony unveiled the Vaio 505 in 1997 with its innovative design in a magnesium body. Most importantly, the Vaio 505 was drastically lighter at 1.35kg and thinner at 22mm when compared to other previously available products. This new design and miniaturization allowed Sony to capture 10 per cent of the laptop market in Japan, a significant share considering that it was based on only one model. Toyota's Prius is another well-known case: It was the first hybrid car, combining gasoline and electric batteries. Its fuel efficiency was two times higher (51 mpg) than the existing gasoline-based automobiles, and it established Toyota as the world leader in this technology. Finally, Canon's inkjet printer, the BJ-10v, came out in 1990. Previous PC printers were relatively expensive at roughly 200,000 yen (approximately US$1,500), therefore less accessible for non-office uses. Canon introduced the BJ-10v at a price of 75,000 yen ($550). Consequently, Canon's market share in inkjet printers soared from 30 per cent in 1990 to more than 80 per cent by 1994. Additionally, this was the first portable inkjet printer and opened up new consumer demand for inexpensive, small printers for home use.

In my research, I used the historical corporate case study method pioneered by Hounshell and Smith (1988). First, I collected data about the three firms from secondary sources: newspapers, business magazines, annual reports, and other corporate reports. Since these were highly regarded firms, there were a number of journalistic and academic articles on the firms, some of which covered product development and organizational dynamics in detail. These secondary sources were valuable when tracing who worked on what aspect of product development, identifying potential interview candidates.

Second, I conducted semistructured interviews with company officials because some detailed information about the process was not readily available from alternate sources. I primarily targeted the project leader of each product, while other project members were also informative. Industry experts, such as stock analysts and management consultants, were included. Interviewing may sometimes be criticized as a subjective process; however, the combination with secondary sources, whose interviews were conducted by other researchers, would strengthen the validity of this research. Interviews added detailed, complementary insights to the study. Although the information I collected was fairly consistent, there were some occasions when people reported different dates and reasons. In such cases, I followed the information provided by the chief engineer or his close associates whom, I believe, had the best oversight of the project.

My interviews were focused on descriptive questions: How did the project start? How many engineers and managers were involved in each stage of the development? Where were they located? What kind of problems did they face and how did they solve? How did the overseas unit contribute in the project? If any distant coordination took place, what was it and how did they communicate? What kind of difficulties did they face? Note that I did not start with questions about geography, why the specific location, eg. Toyota City, was important for the firm or for the project. Such question could embed the idea in the interviewee's mind of providing only proximity-oriented answers from the beginning of an interview. Descriptive questions asking about both local and non-local aspects of a project could allow interviewees to express their own ideas about the project and the location.

Chapters 3–5 will start with a company overview, followed by the firm's internationalization strategy and research development system. The substantial part of each chapter is dedicated to describing the development process in depth. While all development cases are fascinating stories of successes and failures, a simple listing of anecdotal evidence would be long and hard to follow. Therefore, each chapter will be structured on the standard framework provided by product development analyses in business administration literature (Clark and Fujimoto 1991; Clark and Wheelright 1994; Harryson 1998). The development phases are identified as (1) search for technology, (2) concept development, (3) component development, and (4) toward mass production. These are not discrete, mutually exclusive processes, in this absolute order, but interchangeable ones. This framework will allow us to investigate the process more systematically.

Chapter 3
Sony's Vaio 505:
A Portable and Good-Looking Market Niche

> To establish an ideal factory that stresses the spirit of freedom and openmind, and where engineers with sincere motivation can exercise their technological skills to their highest level.
> — Dai Ibuka, the Founder of Sony Objective of Establishment in *Founding Prospectus* (1946)

Sony, a creator of the classic Walkman and the more recent PlayStation, introduced the Vaio 505, a thin, light, stylish laptop computer, into the market in November 1997. This was a design and miniaturizing innovation, and evolved through a series of intense brainstorming exercises for the concept generation, as well as the integration of electric, mechanical, simulation engineering, and materials science for component development. Communication to exchange ideas and provide feedback simultaneously, as well as creating prototypes to evaluate the design and functionality, required physical co-location of engineers and marketing specialists. Therefore, the core part of the development project took place at Sony's technical center in Fujisawa, on the southern edge of the Tokyo metropolitan region, with approximately 20 project members.

3.1 The Sony Corporation

Of the many ways to describe Sony, perhaps two terms would sum up the characteristics of this internationally respected consumer electronics giant: innovation in products and an orientation toward international markets. Sony was the first domestic producer of transistor radios in Japan and the Walkman, a portable radio and cassette tape player. Sony has always pursued the international market. It was one of the first Japanese electronics manufacturers to export to the U.S., to establish overseas sales and production subsidiaries, and to promote non-Japanese to executive positions. This section will briefly examine the history of the company centered upon these two characteristics.

Sony was founded in October 1945 by two prominent businessmen. Ibuka, a talented engineer who was fond of radio-making as his hobby, and Morita, a brilliant salesman later known for his nationalism, started Tokyo Telecommunication Engineering Company with only ten employees in a small room under a department store. In the first four decades, the firm concentrated almost exclusively on consumer electronics, which still makes up the major operational revenue of the firm today. It was the first company to introduce a domestically produced tape recorder in 1950 (Sony 2001, 21, 56). The U.S. firm, Regency, first commercialized a transistor radio in 1954, using transistors from Texas Instruments. In contrast, Sony launched an all internally-produced transistor radio in 1955 (Sony 2001, 115, 118). As exports of this radio successfully penetrated the U.S. and European markets, the company was renamed Sony, which comes from *sonus*, a Latin word meaning "of sound," and *Sonny*, a nickname for a young boy. The rationale was to promote the company as a producer of sound devices in a lively and miniature size (Sony 2001, 134–5).

Unlike other Japanese electronics giants, such as Toshiba, NEC, and Mitsubishi Electric, Sony emerged as a new, not affiliated with *keiretsu*, or business conglomerate player. Consequently, Sony does not represent the Japanese corporate model, as described by political economists (Gerlach 1992). In other words, it does not hold preferential business relationships within the group or strong connections to a main bank, and Sony's procurement is not restricted to its group firms or deeply affiliated suppliers. This main bank system is based on a long-term relationship between a firm and a bank, and the bank provides a stable line of loans at favorable rates, monitors the firm's performance, and holds stocks in return (Aoki 1988; Aoki and Dore 1994). In the 1950s, banks owned only 8 per cent of Sony stocks, while the national average of this ownership ratio was 23 per cent (Sony 2001, 188–9). Their lesser reliance on indirect finance made Sony pursue a strategy of direct finance, including through international bond markets. It became one of the first 16 Japanese firms to issue stocks through American Depository Receipt in 1961, and the first Japanese firm to be listed under the New York Stock Exchange in 1970 (Sony 2001, 434).

Table 3.1   Sales, share, and growth of Sony's business

	Sales in 2000		Sales in 2004		Sales Growth
Division	US$ (mil)	Share	US$ (mil)	Share	From 2000–2004
Electronics	41,471	65.7%	45,754	63.5%	10.3%
Game	5,950	9.4%	7,247	10.1%	21.8%
Music	6,472	10.3%	4,687	6.5%	-27.6%
Movie	4,642	7.4%	7,273	10.1%	56.7%
Financial	3,588	5.7%	5,440	7.5%	51.6%
Other	1,157	1.8%	1,680	2.3%	45.2%
TOTAL	$63,082	100.0%	$72,081	100.0%	14.3%

Source: Sony (2000, 92; 2004b, 90).

Sony has a reputation for its innovativeness, the best examples among its products including a pocket-size tape recorder, Walkman, in 1979, Camcorder in 1983, and PlayStation, a game player, in 1995. Sony's strengths include product development (Itami and Roehl 1987, 102), and it diversified into IT, software, and financial services throughout the 1980s. Although its recent large growth is in the area of games (21.8% since 2000), movies (56.7%), and financial products (51.6%), the bulk of Sony's revenue in the 2000sy still comes from electronics, 63.5 per cent in 2004 (see Table 3.1).

Sony has had a strong orientation toward the international market. It aggressively looked for export opportunities from the 1950s through 1970s, then shifted to production and sales. Sony started exporting early on. Morita, one of the two founders and later CEO, claimed that it was a lesson from the 1950s. Although Sony sold its products nationwide, it particularly relied on the growth of the Kyushu region at that time. This region was heavily dependent on coal mining, and the sudden end of the coal boom in 1955 severely undermined Sony's growth. To strengthen its market portfolio, Sony diversified sales areas into different regions. This diversification included the international market. It was the first Japanese electronics firm to establish direct sales subsidiaries in New York and Switzerland, both in 1960 (Sony 2001, 76, 170, 415).

Morita later launched the strategy of "glocalization" (globalization + localization), (Sony 2001, 457–8). This originally started as a political consideration to lessen trade imbalances between Japan and the U.S. The rising level of exports by Japanese firms when compared to Japan's imports was often a cause for trade friction. Morita foresaw this problem and decided to manufacture in foreign countries rather than export from Japan. In theory, this glocalization meant that Sony as a group would operate at the worldwide level, but Sony in each country would produce and sell products domestically. Sony operated this policy particularly early in the United States, Sony's second largest market. Sony established production facilities for color television in San Diego in 1971, speakers in Pennsylvania in 1974, and videotapes in Alabama in 1977. The overseas operation extended to research and development (R&D), and Sony established its Advanced Video Technology Center in San Jose, California, in 1989. This center was later upgraded to the American Research Lab in 1994 (Sony 2001, 438, 440–41, 461) to include research in semiconductors and digital signal processing (Arimura, 1999, 47). Sony's orientation toward the international market was particularly distinct, in contrast to other Japanese electronics firms. The non-Japanese market produced more than 70 per cent of Sony's sales in 2004, while the same figure was only 20–30 per cent for their rivals (see Table 3.2).

Table 3.2   Sales of major Japanese electronics firms by regions

| | Sales in billions—US$ | | | | | |
	Japan	Overseas	N. America	Europe	Other Area	TOTAL
SONY	21.16	50.73	20.40	16.97	13.36	72.08
Toshiba	36.03	17.09	6.42	4.70	5.97	53.65
Fujitsu	32.48	13.35	3.12	5.82	4.42	45.84
Hitachi	61.20	21.81	7.55	3.89	10.37	83.00
NEC	44.89	9.78	-	-	-	47.18
	Sales share					
SONY	29.4%	70.4%	28.3%	23.5%	18.5%	100.0%
Toshiba	67.2%	31.9%	12.0%	8.8%	11.1%	100.0%
Fujitsu	70.9%	29.1%	6.8%	12.7%	9.6%	100.0%
Hitachi	73.7%	26.3%	9.1%	4.7%	12.5%	100.0%
NEC	95.2%	20.7%	-	-	-	100.0%

Source: Annual report of each company (2004).

Note: US$1 = 104 yen in 2004.

* The total may not be equal to 100 per cent due to intra-company transactions

Sony's international orientation extended not only to production, but also to incorporation of non-Japanese personnel. Japanese multinational firms often used Japanese expatriates for the management of overseas subsidiaries (Lam 2003, 697). However, Sony appointed an American as chief of U.S. sales operations as early as 1972, and listed non-Japanese on the executive board even for the Japanese headquarters in 1989 (Sony 2001, 413, 447). The use of non-Japanese personnel was highlighted when the chairman, Idei, announced in March 2005 that Sir Howard Stringer, a Welsh-born businessman, would be the new chief executive officer (CEO) of the entire group (Sony 2005d). This surprised both business and the public because non-Japanese had, historically, rarely led major Japanese firms, with the exception being Carlos Ghosn, at Nissan since 1999. The financially troubled Nissan gave up its shares to French Renault, and successfully turned around within two years. Sony, similarly experiencing difficulties, faced "Sony Shock" and a 25 per cent fall in its stock price in 2003. However, the shock was never as severe as that of Nissan. Sony claimed that the appointment of Stringer, who had previously led Sony Entertainment and served as a board member of the Sony Group, was not "a surprise for the global company," and would continue the integration of media and IT technology (Idei, cited in *New York Times* 2005).

3.2 Sony's R&D System

Sony spent roughly $4.9 billion on R&D activities in 2004, which was 7.5 per cent of sales. This ratio was considerably higher than that of the industrial average within the electric machinery sector, 5.0 per cent. Their electronics sector received most (83.5 per cent) of this spending (Sony 2004b, 25, 87). Organizationally, these R&D activities take place at four levels: (1) the Research Center, (2) corporate laboratories, (3) various development divisions of each business unit, and (4) overseas R&D centers. First, the Research Center, located in Atsugi, Kanagawa, one hour west of central Tokyo, conducts medium to long-term basic research. This Center employs 250 researchers, including physicists, chemists, electrical, and mechanical engineers. Since the 1980s, the Center has recruited foreign scientists, and recently counted more than 60 foreign employees working on a one-year contract basis, with possibly unlimited renewal (Sony 2001, 372). This Center was under direct control of the board members and made bi-annual financial reports. The board concerned itself more with the budget than the content of projects (Harryson 1998, 127). Therefore, the Center enjoys a freedom to choose which types of research projects it engages in.

Second, the corporate laboratories aim to bridge the gap between the technology of the Research Center and the final commercial products in the short to medium term. Although Sony has notoriously and repeatedly changed its organizational structure since the early 1990s (1994, 1999, and 2002), these changes did not significantly affect R&D. Essentially, researchers moved back and forth between organizations. There are approximately 600 engineers in mechanics, electronics, optics, and sound processing, located at the headquarters buildings in Tokyo. In the mid 2000s, the 11 corporate laboratories cover a wide range of technological areas, such as telecommunication, Internet-related application software, and advanced computer devices (see Table 3.3).

Table 3.3   Sony corporate laboratories

Lab Name	Research Area
Contents and Application Lab	Entertainment
Broadband Application Lab	Broadband, application architecture technology
Network CE Development Lab	Audio visual equipment and applications
Ubiquitous Technology Lab	Telecommunication system, security system
Storage Technology Lab	Storage devices
Display Technology Lab	Optoelectronics
Material Lab	New materials
A3 (A-cubed) Lab	Business strategy
Digital Creatures Lab	Robotics
Integration Area Lab	Molecular materials, electronics
Cyber Technology Lab	Intellectual, IT technology

Source: Sony (2002, 48–9).

The third organizational tier comprises the development divisions in each business unit, with approximately 8,000 engineers in total (Tambata, interview, September 24, 2004). The orientation of these development divisions is to create commercially viable products from innovative ideas in a short space of time. The development of the Vaio 505 took place at this level of the R&D system. IT and Mobile Solutions Network Company, one of Sony's 11 business units, housed IT Company Division II, which was in sole charge of the development of the Vaio notebook (Ito, interview, September 21, 2004; Sony 2005c).

Lastly, Sony has nine overseas R&D centers in the U.S., U.K., Germany, Singapore, and Belgium. These laboratories have in total approximately 500 researchers. The earlier mission of the overseas labs was (1) to modify exported products for local markets, (2) to find advanced technologies in the local market, and (3) to provide technological support for overseas plants. Later, their mission shifted more toward (4) advancing their own technology and creating new businesses for local needs. Although it is beyond the scope of this section, the effectiveness of these four missions is in question, especially the last, creating new businesses for local needs. This is not a simple process. Furthermore, some studies (Arimura, 1999, 48–9) pointed out that Sony had not had successful collaboration at the cross-border level or in synergistic research.

Table 3.4   Sony's overseas R&D laboratories

Name	City	State	Country	Year
Intelligent systems research lab	San Jose	CA	U.S.A.	1991
Research lab	San Diego	CA	U.S.A.	1991
Research lab	Boulder	CO	U.S.A.	
Research lab	Montvale	NJ	U.S.A.	1991
Precision engineering center	Singapore		Singapore	1987
Stuttgart technology center	Stuttgart		Germany	1985
Technology division	Basingstoke		U.K.	1978
Digital communication division	Pencoed		U.K.	1992
Telecom Europe			Belgium	1990

Source: Arimura (1999, 47).

3.3 Sony in the PC Industry

Vaio was not the first PC created by Sony. The history of PC production at Sony goes back a few decades, and shows both Sony's ambition in the PC industry and the repeated failures it faced. In sum, it took three decades for Sony to launch its first widely recognized computer model in the market.

Sony first produced a transistor-based calculator, SOBAX, in 1967 but eventually pulled out of the business because of fierce price competition from Sharp and Casio (Kawaguchi 2003, 66). In 1982, it collaborated with Panasonic and introduced the SMC-777, a computer which was based on the MSX standard developed by Microsoft and which had extended graphics-designing functions. Incidentally, the leader of this SMC project was Idei, the former CEO who aggressively launched the Vaio business after 1996 (Tateishi 2002, 276). Other non-commercial ideas include NEWS, a workstation, in 1987, AX in 1988, and IBM-compatible QuarterL in 1993 (Sony-EMCS 2005). Aside from producing PCs in its own name, Sony stayed in the industry by being a supplier. Sony had original equipment manufacturing (OEM) contracts with Dell and Apple from 1991, and it was a long-standing wish to produce PCs under its own brand.

Advancement of information technology in the mid 1990s brought another business chance for Sony. In addition to office use, PCs started to penetrate household markets both in Japan and in the U.S. PCs at home meant further variety in uses. In addition to traditional word processing, people started to use PCs for graphics (picture and movie), music, and games. This opened up the possibility of audiovisual use for PCs, and Sony considered this as its next primary market and technological need. VAIO is an abbreviation of Video Audio Integrated Operation, and explains Sony's ambition to unite PCs, telecommunication, music, and movies. Idei, who took the chairmanship in 1995, appointed Ando, who championed the games industry with PlayStation, to succeed him at the IT Company. The high executive level started to discuss entry into the PC market in June 1995, and Idei publicly announced Sony's intention to enter the market by the end of that summer (Tateishi 2002, 277).

Sony started to produce its first Vaio series in desktop, MiniTower, and notebook forms in July 1997. As a first step to integrate the PC, audio and visual use, the newly introduced desktop had music CD players that operated at double speed and software to download pictures from Sony's digital cameras. It was introduced in the U.S. as early as it was in Japan. However, with a price of $3,000 to $5,000, Sony could capture only 1 per cent of the home PC market despite spending $20 million on advertising. Michael Dell dismissed it as "the invisible invasion" (cited in McWilliams 1997).[1] A PC with a few new music and audio functions was not appealing to consumers, and at this stage Sony could not distinguish its PCs with the Vaio orientation. The challenge was enormous for Sony, and it had to seek different business models to stand out in the PC market.

---

1   Sony was not the only loser in the U.S. PC market at this time. Hitachi and Fujitsu spent $30 million and $25 million, respectively, on advertising, but captured less than two per cent market share. The exceptions were Packard Bell NEC for desktops and Toshiba for laptops.

3.4 Development of the Sony Vaio 505

The development of the Sony Vaio took place essentially at one site, Fujisawa, Sony's technical center located 45 km southwest of its headquarters in Tokyo. A ten-member team started to develop the product concept in November 1996, increasing to 16 members by the end of component development in summer 1997. Once the process reached production in July 1997, the geography started to diffuse to Nagano, three hours northwest of Tokyo.

The concentration of development activity in the earlier phase was a result of heavy interactions between 11 project members: five electrical engineers, three mechanical engineers, two software engineers, and one designer. Throughout the first two key development phases, concept development and technological development, project members benefited from geographical proximity. First, toward the beginning of the project, the project members carried out a series of brainstorming exercises to generate concepts of the new product. This was a heavily face-to-face communication-based process conducted in one designated room. Then, in the middle of the project, they had to integrate various types of technologies, such as mechanical, electrical, materials, simulation engineering, and art design. The objective of this integration was to achieve maximum miniaturization of the laptop, which was one of the key early concepts. Both development activities required sharing the most up-to-date knowledge among the core members, and repeated trials and errors in redesigning. Engineers had to be physically present at one location to see, touch, and test prototypes. In essence, the development of the Vaio required intra-divisional coordination of the core members, between the staff at the Design Division I of Sony IT Company and a design center at the headquarters, and co-location of all these deeply involved members was necessary.

At the same time, coordination with other groups could not be neglected. The executive board located in Tokyo was the key to initiating these new product developments, as well as continuously supporting the project through funding and connecting crucial personnel within Sony. Its marketing division, also located in Tokyo, was important to provide feedback and to a bridge between development and sales staff. There were a few core suppliers in the Tokyo region involved in designing core components. Last but not least, a production arm, Sony Engineering, Manufacturing, and Customer Service (EMCS), tightly coordinated with the design team to shift to mass production. The development process mainly took place in Tokyo, then spread to Nagano for the production process, but stayed within a three-hour journey from headquarters.

3.4.1 Concept Development

At this early stage of new product development, the project members were ten engineers located at Fujisawa. The coordination was almost exclusively within one group, and exchanges of ideas through face-to-face communication were essential to generate product concepts.

The project leader for the Vaio 505, Susumu Ito, joined Sony in 1982, and was involved in the development of earlier PCs. As Sony exited the market each time yet maintained its ambition to re-enter the computer market, Sony assigned him to continue to work on the development of peripheral devices for PCs. From 1994, Ito was stationed in Mountain View, California, and worked on the development of mobile terminal devices in collaboration with Apple, Motorola, and AT&T. At the start of 1996, the Sony executive called him back to Japan to develop a new PC under the Sony name.

He and nine other engineers worked on the development of a laptop from February 1996. They designed and completed a prototype of an A4-size model (236 x 297 mm), the standard laptop size of the period. However, they decided to terminate production by October 1996. Sony systematically reviewed the market feasibility of a potential product in comparison with competitors; this process is called a line-up meeting. At the meeting, they analyzed products from Toshiba and IBM, and could not find any originality in and positioning of the Sony product. Consequently, they dissolved the project team, but five members remained to search for an alternative laptop computer.

Those five remaining members, four electrical engineers and one software engineer, were affiliated to Design Division I of Sony IT Company (Goda 1999). This development division was located in Sony's Shonan Technical Center at Fujisawa, 45 km (28 miles) southwest of its headquarters in Central Tokyo. The project members started to work on concept development of a new laptop model in November 1996, and completed it by the end of February 1997.

At that time, the executive board ordered the team to create a laptop unique to the market, i.e. something not similarly produced by their competitors. The team went through another line-up meeting, at which they discussed the types, prices, and functions of competitors' products. Here, DRAM and CPU were two examples of the features discussed. Although Sony had the capacity to produce DRAM, the team decided not to compete over the capacity of this component with other companies. Past experiences in the DRAM war had taught them that the fastest PC with the biggest DRAM would still be obsolete in just six months. Additionally, with competition from Korean and Taiwanese DRAM producers, competition over speed where DRAM was concerned would be unprofitable. It was impractical for Sony to produce its own processing unit. Sony had to use Intel's CPU and Microsoft Windows. As soon as they decided to enter the PC market, the chairman Idei started to negotiate with these two U.S. giants to procure those components.

In the end, the Vaio team came to the conclusion that they would not produce a laptop intended to compete in terms of larger or faster capacities (Ito, interview, September 21, 2004). Instead, they searched for a different type of competition. Toshiba and IBM produced laptops, but these were still relatively expensive, normally $3,000 or higher, and marketed mostly for office use. The design of these models was classic, black, bulky, and physically portable but not easily, nor desirable. Ito calculated that there would be a market for non-office use and launched the concept of the new model, "portable and good-looking in arms."

Figure 3.1   Sony's Vaio 505

Source: Author.

More specifically, it would be smaller, thinner, lighter, and with a different colored exterior body. Based on this concept, the team started to design the body of a laptop. Instead of the conventional A4 size (236 x 297 mm), they decided to scale it down to B5 size (208 x 259 mm). They figured that scaling down to notebook size would substantially improve the portability of the PC. Ito proposed the thickness be limited to 23 mm, in contrast to the previously available 37.6 mm, and the weight limited to 1.35 kg, instead of 2.4 kg (Sony 1997). In contrast to the normal dark-colored, plastic exterior, they decided on a shiny, magnesium body surrounding all six dimensions of the laptop.

This concept of a slim and light laptop came not only from the five original project engineers, but also from a designer, Teiyu Goto, who participated and contributed substantially in this process of concept development. Sony had a two-tier system of design organizations: a product design division in each business unit and the Creative Center answering to the headquarters (Tambata, interview, September 24, 2004). Goto, affiliated with the Creative Center, was a veteran designer and had previously designed the famous PlayStation. As friends, he and Ito informally started to discuss the concept for the new model as early as November 1996. By the end of 1996, Ito and his section chief, Tanaka, officially requested that Goto be a part of the project instead of designers from the business unit. Goto was stationed at Shinagawa, Sony's headquarters in central Tokyo, and was working on the design of the PlayStation 2 at the time. This PlayStation was what Sony called a total design coordination product (Goda 1999) wherein, after an internal design bid, Sony selected one designer and designated one full year for designing that product only. Moreover, the designer would stay with the PlayStation project and be involved in designing future generations. Therefore, it would not be possible to have Goto entirely committed to another project. However, since the Vaio 505 was a high priority project and Goto had a personal connection and commitment with Ito, this project turned out to be an exception, and Goto worked simultaneously on the two. There was no official transfer involved, but Goto visited Fujisawa at least once a week to participate in the concept and design discussion.

Information technology helped Goto to join and keep up with the Vaio 505 project. Although Goto and Ito had informal communication, Goto needed to become familiar with more detailed debates in the evolution of the Vaio 505 concept: Understanding earlier ideas was critical to producing shared ideas for the future. Although the concept of the new product could be described in one phrase, i.e. portable and good-looking in arms, the original project members shared a tacit understanding of this concept and simply knowing the final result was not sufficient. More specifically, it was important for Goto, first, to understand what other concepts had been debated and rejected. Then, Goto could share the concept more deeply by going through what exactly "portable" and "good-looking" meant to the other members of the group.

The project members produced meeting memos and posted them on the intranet so that every team member, including Goto, was constantly updated. Sony's intranet system allowed Goto a voice even before officially joining the

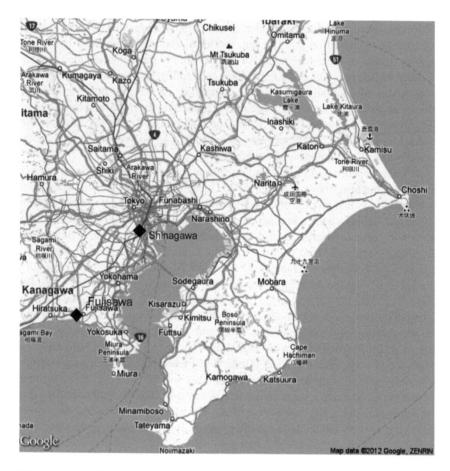

Figure 3.2   Map of Vaio 505 development

Source: Author.

team (Goda 1999). For example, he commented from an early stage that clock frequency of CPU would not matter, and that the slimness of the final product would be the greatest attraction to consumers.

In sum, what was used was a hybrid of two modes of interaction: physical visits by Goto to have face-to-face conversations with the rest of the team, as well as information updates with the main group and some input by Goto via information technology, namely intranet and e-mail. Both modes were necessary as a visit by Goto once a week was not sufficient to transmit all the necessary information. E-mail provided detailed and updated information on at least day-by-day basis, and this use of both modes was effective. On the other hand, it is important to note that there was a clearly hierarchical relationship between the modes. The absence of information technology would have caused inefficiency in communication, but could have been replaced by more frequent visits, say three to five times a week. This would certainly have decreased Goto's flexibility and time commitment to other projects. Yet it would have been possible in a strict sense. In contrast, the same condition did not hold for the other side. Goto might have increased the use of information technology, perhaps by including telephone calls, but that in itself would not be sufficient. Face-to-face communication was still needed, and was a more important mode. Thus, information technology facilitated the development process, but did not replace it.

3.4.2 Component Development

Component development required coordination not only within Design Division I, but also with other divisions, such as a designer from the headquarters and a few key suppliers. All those involved were located in the Tokyo region, more specifically the Tokyo-Yokohama-Fujisawa corridor.

Once the project team confirmed the concept, they started to develop components in February 1997. In addition to the original six members, two electrical and one mechanical engineers joined the team. Furthermore, a new director was assigned to the division at Fujisawa. These four people had previously worked on the designing and development of camcorders, and this was a critical addition for the Vaio 505 (Ito, interview, September 21, 2004). As the name *Vaio* states, it had to integrate audio and visual (AV) functions into PCs. This integration would include the use of music and audio files in a digitized format (software), as well as the direct link to a camcorder (hardware). Development of a laptop with the assistance of AV-oriented engineers was indispensable. Additionally, the audio staff were highly experienced in designing from a consumer's point of view, in contrast to Ito and the other original members, who were rather classic electrical engineers. They were aware that the limitations on a stylish and functional model as imagined by engineers could be different from what consumers expected. This was an effort by Sony to incorporate as much market-forward feedback as possible even in the design and production phase.

February to March 1997 saw the peak of the brainstorming meetings. They repeatedly drew designs, created prototypes made of paper, discussed, scratched it all, and started over. They designated one meeting room for this project, where they placed updated designs and prototypes, and there were almost always some project members in this room discussing the sophistication and stylishness of the design, its use, and functions. These discussions had to be face-to-face and involve a prototype, otherwise some points could not be clearly discussed or effectively compared. It was almost impossible to rely solely on oral, non-visible communication for the arrangement of components, especially where they should be placed. Everything would have to come with a precise visual explanation so communication via telephone at this stage was undesirable, because it could create confusion.

Furthermore, the information had to be shared by all project members in real time—a model design from even the previous day was obsolete. Time was always precious, and loss of it for catch-up meetings would be a waste. More importantly, people's ideas would not flow if interruptions occurred, and they had to minimize unnecessary interruptions (Ito, interview, September 21, 2004). For example, if one member missed a meeting and had to ask why and how the rest of them came up with a new idea, that would require a review of previous information, and delay producing ideas for the next step. In sum, the brainstorming exercises required that all project members to be present at all times. Earlier in the project, the designer Goto had come to the Technical Center two to three times per week on average. However, from February to early March, at its peak brainstorming period, he came almost every day (Goto, interview, September 22, 2004).

While they consolidated the design, the project members started a search for components for which they had a couple of criteria. The first was for components to be standardized and mass produced. In other words, they searched for components that were inexpensive and readily available. Second, it was desirable that a component should be produced by Sony or its affiliates, but this was not a strict condition. If a component was unavailable at Sony or available but significantly more expensive than purchasing it from the market, they did not hesitate to buy from suppliers (Ito, interview, September 21, 2004). One exception was the exterior body, made of magnesium—Sony's first attempt to produce such a slim shape in this material. They considered this skin to be a critical part of the product and did not want to outsource, but it turned out that the cost of producing the exterior was considerably higher than expected. Then, they had to consult the executive board immediately to see whether it was feasible. Eventually, Ito and the executive board agreed that economies of scale would likely solve this cost issue. That is, if the Vaio 505 were to sell as well as expected, the cost per laptop would decrease sufficiently. Therefore, it was not only the proximity between engineers, but also the proximity between the engineering and executive divisions that was essential.

In relationships with suppliers, Sony was willing to collaborate and even to invest. For the modem connection, the team looked for a thinner connecting device than previously used. The conventional size for a modem connector was more than 25 mm square. However, as the thickness of the Vaio 505 was set as 23 mm, they had to shrink the telecommunication device. The solution proposed by Sony was to eliminate plastics in the connection between the top and bottom of the rim of the connector. The plastic on the left and right sides would still stably support the connection. This component had been previously procured from a supplier, and now they had to make changes in their production line, so Sony invested approximately 200 million yen ($2 million) in the supplier.

The collaboration with suppliers extended to a level whereby it involved working with a competitor in the industry. The Vaio 505 monitor was liquid crystal display (LCD) and supplied by Toshiba, Sony's main rival over laptops. LCDs were becoming popular in the late 1990s and were widely used in laptops. However, the market demand was going for a larger size, such as for TVs. In contrast, the Vaio 505 required a smaller size, 10.41 inches, in lieu of the conventional 12-inch or 14-inch monitor. In addition to size, Sony considered it necessary to have a monitor with higher resolution. The Vaio was intended to integrate PCs and multimedia, and its expected uses included watching movies, graphic design, and games. Therefore, the resolution conventionally available for LCDs was not sufficient, and Sony requested that Toshiba develop a new model. Toshiba responded willingly, and supplied a smaller, yet higher resolution monitor to its competitor. The Toshiba Lab was located in Kawasaki, within an hour of Fujisawa.

These close, modifying processes of component procurement show that relationships with suppliers were often complex. In other words, it was not as simple or static as Sony telling their suppliers what to make, and those suppliers providing the specified components. Instead, the relationships were dynamic as Sony and its suppliers continued to discuss, produce, modify, and re-evaluate components. Under such dynamic circumstances, it was more convenient to use suppliers in close proximity (Sakaguchi, interview, July 21, 2004).

Designing and allocation of components was not a simple spatial matter, but required functional and other miscellaneous coordination. Often, the team discovered these allocation problems only after designing. Then, they would look for a solution, which could require redesigning of the component. This whole

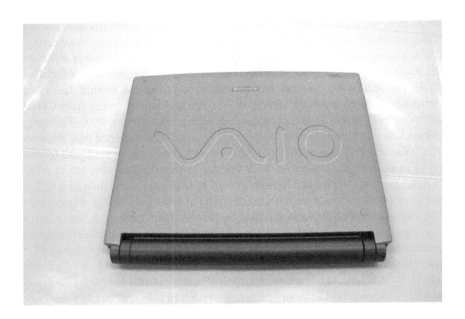

Figure 3.3   Backside of Vaio 505

Source: Author.

process was very much trial and error. "Learning by doing" (Irwin and Klenow, 1994; Jovanovic and Nyarko, 1996; Young 1991) is a concept in organizational studies for making innovation. The design phase of the Vaio 505 could be more precisely described as "learning only after making mistakes."

The arrangement of components required not only the technique of miniature engineering, but also the art of creativity. Most laptops were equipped with a lithium battery, which was conventionally a cylinder shape, 20 mm wide. With the new Vaio's thickness limit of 23 mm, the location of this battery turned out to be a major issue. A laptop must have a thick enough keyboard (about 5 mm), yet that would leave less than 18 mm for the battery space. The non-keyboard space must be allocated to other key components on a PC: a motherboard with CPU, video slots, and a hard disk drive. The Vaio team considered requesting development of a new thin battery from the battery division, in which Sony has had market competiveness. However, they realized this development would be substantially costly in terms of time and budget.

Interestingly, the solution came not from the engineers but from the designer. Goto proposed that, contrary to the conventional wisdom, the battery did not need to be located under the main body of the PC. Instead, it could be placed at the junction of the main body and the monitor. The conjoining part could be a cylinder shape and could be as thick as both the main body and monitor combined. Thus, the battery could be located inside the conjoining part. Ito immediately accepted the idea, and the Vaio 505 got a distinct cylinder joint at its spine (see Figure 3.3).

Another battle over space was the balance between the miniaturization and the management of technologies related to thermodynamics and software engineering. It was a combination of several types of engineering technologies that achieved the small scale of the Vaio 505. The volume of the Vaio 505 was reduced to 40–60 per cent of the previous model, which meant that the heating density of components soared inside the laptop. In PCs, the heat mainly comes from CPUs, and a processor with 133 MHz could increase the temperature by 49°C. Similarly, system LSI would increase the temperature by over 50°C and hard disc drive by 12°C (Koyanagawa et al. 2000, 10). Moreover, the engineering team had to step up to an even bigger challenge. They decided to remove a cooling fan in order to make the body lighter and thinner, as well as to present the feeling of a higher quality computer without the fan noise. In order to solve these heat problems, they first had to create effective aerodynamics. In what Sony called the "progressive cooling system," they carefully designed ventilation holes so that air would be efficiently taken into and out of the laptop without a fan. Second, they set up heat pipes which were low in conductivity and could transport heated air efficiently outside the machine. However, the insertion of heat pipes affected the arrangement of other components. The team had to test which arrangements would work most efficiently.

The team used a three-dimensional computer aided design (CAD) program to estimate the heat transmission process (Iguchi, interview, December 4, 2004).

Nonetheless, this redesigning for component arrangement was a lengthy trial and error process.

> With a variety of choices in materials for the body, methods to strengthen body structure, and location of components, it was a process of creating a prototype, abandoning it, and re-creating it a few dozen times. Some days, I was feeling that we were finally getting close to the optimal body structure and device location. However, people [other project members] kept providing me new information, such as that there was a better material available, or a request to improve cooling efficiency, etc. I had to abandon my designing again, again, and again. (Laugh.) I tried to be optimistic, and told myself repeatedly that we could achieve a better and lighter laptop. (Asawa, cited in Sony 2004a)

> Asawa-san [the mechanical design leader] and I coordinated all the time. When the design of the body changed, we re-designed the [mother]board. After the board adjustment, we re-made the body. It was a very close and lengthy coordination. (Ishikawa, cited in Sony 2004a)

This specific type of innovation, miniaturization, involved both electrical and mechanical engineering, as well as materials science. It was a trial and error, or more precisely, a trial and more trial process. Additionally, a change in one component caused snowballing changes in other components. Focusing upon the two mechanical design engineers, this was coordination in real time.

Information technology was critical to the design process, but did not always provide solutions. Human creativity superseded the efficiency of machines, and higher density designing came from rule-breaking. In CAD, there were rules that every registered component needed a certain space between it and other components, determined by component suppliers. The suppliers determined the necessary space based on the standard use of a component. In reality, using the entire specified space was not as necessary as suppliers had suggested, and the Vaio engineers designed with an intentional overlap in the spaces. This overlap could reduce un-used space, and therefore could achieve higher density. However, it required close examination and hands-on placement of every component (Miyano, interview, December 4, 2004). A mechanical engineer checked that each overlap would not cause any significant dysfunction. It was an extremely tedious and lengthy process to go over every one of 1,100 components in the Vaio, but it proved to be effective.

To summarize the component development phase, four factors required co-location of the players involved. First, converting the concept to actual component allocation required testing with physical prototypes, conducted by engineers. This meant co-location of the engineers involved and face-to-face communication. Second, a dynamic process with suppliers denotes physically close relationships with them. Third, the design innovation took place through an integration of multiple fields of knowledge, more specifically electrical engineering and artistic ideas, seen in the successful example of the re-positioning of the battery space. Lastly, the need for constant adjustment, such as component allocation, necessitated all involved engineers being located close to one-another so that they could readily access updated materials and design information, as well as notify others of changes.

3.4.3 Toward Mass Production

At this stage, the geography of the participating players went from Fujisawa and Shinagawa to Nagano, a production site (see Figure 3.2). The period of the most intense face-to-face communication was over and not all of the project members had to meet every day. Broader coordination between the development and production divisions was needed.

Coordination took place not only within the project group, but also between different divisions. From April 1997, they started to hold a project progress meeting every month. To this meeting were invited key personnel from six divisions: design, material, product quality, applied technology, software application, and production. They met in Shinagawa, the headquarters and the location of four of the divisions: materials, product quality, applied technology, and software application. The production division was located in Nagano, a three-hour journey from the headquarters, but the personnel commuted to every meeting. The executive board did not get involved in this meeting directly. However, Ito separately made a progress report to the executives by sending a meeting memo. "It was important to continually update the progress to both the executive board and related divisions. We [the design team] of course communicated with a specific division on an ad-hoc, as needed basis, but the monthly meeting facilitated our coordination process" (Ito, interview, September 21, 2004).

Production involved suppliers, and preparation for the whole coordination would normally take four months for PCs (Komamura, cited in Sony-EMCS 2004a). In order to launch sales in November 1997, Sony started to bridge design and mass production in July. Sony outsourced its Vaio 505 production entirely to Sony EMCS, its 100 per cent subsidiary, at its Nagano site, three hours northwest of the headquarters. EMCS stands for Engineering, Manufacturing, and Customer Service; it was Sony's strategy to separate corporate functions into two organizations. Sony IT Company concentrated on product development, product design, marketing, and sales, while Sony EMCS focused on production design, mass production, and customer services (Sony-EMCS 2005).

Toward the end of the development process, more configurations for software applications were needed. At the core location, Fujisawa, an additional 30 software engineers joined the team, while the number of electrical and mechanical engineers did not increase. Like any PC, the Vaio came out with a series of software programs, including the operating system (OS) and Internet connections. To incorporate audio and visual dimensions, Vaio 505 additionally contained graphics, map navigation, sound and multi-media audio programs—a total of 33 software programs (Sony 1997). The additional software engineers worked on the development, coordination, and adaptation of software programs. Although 30 engineers were involved, most of them were also involved in the software development of other components in the Vaio series. Therefore, it was about four to five engineers who were mainly committed to the Vaio 505 model (Ito, interview, September 21, 2004).

Sony bridged design and production on two organizational levels. First, more conventionally, the production engineers from both factory and design sides closely coordinated, exchanged information, and confirmed each process. Second, rather uniquely among Japanese firms, they appointed the project leader of the designing division, Ito, to also be the leader of the production group at the factory.

A manufacturer always has to coordinate design and production. This coordination requires sharing of knowledge on both sides, and conventional organizational practice includes the exchange of personnel between the technical center and the production line. "Prototype making is always different from mass production, and the job of design engineers at the factory is to narrow the gap between the two. From the production side, on average, I made business trips [to the designing division] three times per week" (Suyama, cited in Sony-EMCS 2004b). The production engineers from the factory had two missions: First, as stated, they had to confirm that what was specified by the design team could be produced on the production line. Additionally, they had to coordinate production processes, including procurement from suppliers. The factory had to achieve the procurement on time and with the expected quality.

The higher level of miniaturization in the Vaio 505 meant higher precision in locating components and a more careful production process. This precision was particularly important for the smaller components. The design engineer carefully tested the production line at the factory several times to confirm that it could produce exactly what the design team wanted (Miyano, interview, December 4, 2004). Excluding the design prototypes, the formal production prototype was produced twice: once to test mold-making and the other to test mass production (Ito, interview, September 21, 2004).

The second organizational method Sony used to bridge development and production was the appointment of Ito as the head of the production division. Previously as a project leader, Susumu Ito held the title of deputy general manager in Division II of Sony IT Company. To oversee the production and bridge two divisions, Ito was also appointed as a senior general manager of the Vaio Design Division of Sony-EMCS. This was a formal yet flexible use of organizations at the Sony Group level. Sony did not have to formally divide the divisions or the personnel between the design and production units but with the same person responsible for both the units. Ito was then based in two locations and spent half the week at the design division and the other half at Nagano.

There was collaboration with overseas R&D laboratories, from LSI Systems Lab of San Jose and Telecom Europe of Brussels, however, their role was limited. As the development took place in Japan, the Vaio 505 interface was, by default, in Japanese. Although the computer programming language was universal, all the screen messages appeared in Japanese. Thus, the role of overseas labs was local adaptation, but only as a marginal translation of the brochure, PC labels, and configuration of software programs. Local adaptation could potentially cover a wide range, such as adding new functions, changing exterior design, etc. but the Vaio team did not consider these possibilities, deciding that the model would be universal except for the interface. Therefore, there was no extra role expected for overseas R&D laboratories.

3.4.4 Mass Production

Mass production started in October 1997, a month before it was introduced to the market. By this stage, most of the design and software engineers at Fujisawa were no longer actively involved, and the major players were the production line staff in Nagano. Of the 11 factories of Sony-EMCS, the Nagano site was designated to produce Vaio laptops and AIBO, a robot dog (Sony 2004c, 3). An estimated 1,550 employees were at this site, and about half of them worked on the production of various Vaio units.

The final intra-firm coordination was between the design team and the marketing division. In 1997, Sony delegated the marketing function to its subsidiary, Sony Marketing. It was in charge of marketing and sales of all Sony products in Japan. Based in Shinagawa, the location of Sony headquarters, the marketing arm had about 3,600 employees (Sony-Marketing 2005). Ito presented the product concept and targeted consumers for the Vaio 505 to the marketing staff. Goto, the chief designer, explained the theme of the exterior design. This sharing of the product image was critical for the sales staff to establish an effective sales presentation (Goto, interview, September 22, 2004).

3.5 Summary

Simply stated, the critical development of the Vaio 505 took place at one site, Fujisawa, where Sony's Technical Center and IT Company Division II were located. A key designer from the headquarters commuted extensively to Fujisawa to discuss the concepts and design development. Additionally, major component suppliers were also located in the Tokyo region. Only from the point of preparing for mass production did the geography of production open up as far as Nagano. There was a role for overseas R&D laboratories, however, it was limited only to translation.

The Vaio development case stresses two critical aspects of innovation making: (a) prototyping and testing, and (b) the interdisciplinary nature of technological advancement. First, engineers and designers shared updated design information, provided feedback, and made further model changes by creating prototypes. They started with the general concept of the product, that it should be portable and good-looking when carried. However, this was not the end of the design process, but only a means to reach the end. Each member still had to figure out what was meant by "portable and good-looking" and how that could be expressed in the actual laptop. In a way, this prototyping and testing was a process of generating and configuring tacit knowledge among the project members. They all knew the general direction of the final product, but not exactly *what* it would turn out like. For this process, sharing the norms and routines and having face-to-face discussions was not sufficient, but required physical colocation of the project members, as well as active discussion using prototypes. They created hundreds of prototypes with paper and polystyrene, designated one meeting room for the project as a place for the latest model, and discussed the design repeatedly. Sony's Technical Center at Fujisawa was the place for this project process.

Second, innovations often took place within an interdisciplinary approach. The best example was when the Vaio team faced limited space for the battery: the designer solved the problem. It was a simple method to remove the battery from the conventional space in the main body and place it between the main body and the monitor in a distinct cylinder shape. Here, adding an idea from the creative design perspective was the most efficient and effective solution to a problem that was conventionally solved within the field of chemical engineering. Additionally, when the Vaio team decided to remove a cooling fan, a mechanical component, they replaced the function by introducing new materials, the field of chemical physics, and configuring the most optimal aerodynamics via simulation technology. Electrical, mechanical, software, and design engineers collaborated to tackle these issues.

This shows that innovation did not necessarily happen on a single trajectory. In other words, a better solution did not necessarily come from the higher technology. An electrical engineering problem was not necessarily solved by gathering together more sophisticated electrical engineers. Solving the technical challenge was not linear, but could take place in multiple directions. Therefore, it was the availability, interaction, and management of several technical fields that shaped the new product. Sony had to manage a pool of highly skilled labor in electrical, mechanical, and simulation engineering, as well as chemistry and physics. This interaction was most effectively possible in close proximity, so that the engineers and researchers could freely exchange, test, evaluate their knowledge, and create anew.

Chapter 4

Toyota's Prius: Call for Next Century Induced Hybrid Technology

Always think about what's next from where you are standing. The true professional is someone who can improve further out of the improvement you have just made.

— Taiichi Ono, former Vice President of Toyota, the legendary founder of Toyota Production, cited by Fujio Cho (2001), former president of Toyota

While Toyota is best known for its efficient production system and product quality, this chapter will discuss the story of the Prius. This giant of lean production created the Prius, a car with hybrid technology—gasoline engine and batteries—in 1997. The Prius attracted widespread attention from both industry rivals and consumers due to its high fuel efficiency, hence Toyota marketed it as an environmentally friendly car for the 21st century. The product development of the Prius essentially took place in the core Toyota region, Toyota City and Higashi Fuji, two hours east of Toyota City. With its complexity to develop, coordinate, and assemble, Toyota had a constant struggle among key component development divisions and production facilities. The process of concept development required core members to be able to exchange ideas intensely and at one location. The hybrid system had three major subsystems: the batteries, cooling system, and overall system integration. They therefore had to coordinate tightly between the three divisions. Furthermore, the overall organizational coordination stretched to more than ten units among development, production technology, assembly divisions, and suppliers. Thus, Toyota had to strategically locate all the involved units in geographical proximity for this intense intra-firm and inter-firm coordination.

4.1 The Toyota Corporation

Toyota today is known for its production system and wide international sales, particularly in the U.S. From the beginning, the company emphasized high quality, just-in-time production systems and continuous process improvement. On the other hand, it was not an early player in the international market. Only after the late 1980s, did it start to pursue sales and production in the U.S.

The history of Toyota dates back to the 19th century. This giant of the automobile industry, a symbol of mass production and technology in the 20th century, incidentally has its roots in the industry and technology of the previous century, textiles. Sakichi Toyota, a famous philosophical thinker and inventor, created a manual wooden loom in 1890, and later introduced a series of power looms, including a continuous automatic model in 1924 (Toyota 2005a). Based on his philosophy of through inventions and contribution to society, Sakichi desired his son, Kiichiro Toyota, to explore a new field for a new era—an automobile, an untouched industry in Japan at the time (Reingold 1999). Kiichiro successfully created the first domestically-produced automobile, A1, in 1935. Soon, Toyota Automatic Loom Works, still a parent company of Toyota Group today, spun off its automobile division, which became Toyota Motors in 1937 (Toyota 2005a).

In 2008, Toyota became the largest automobile manufacturer in the world by surpassing General Motors (GM). It has sold cars in more than 140 countries, totaling $166 billion in sales annually. This produced a record annual profit of $15.8 billion, larger than that of GM, Ford, and Daimler-Chrysler combined. With its four brands, Toyota, Lexus, Daihatsu, and Hino, Toyota employs 260,000 staff in a consolidated base (Toyota 2004b, 1–2). Its market share in Japan is 42.9 per cent, while the same figure in the U.S. is 10.5 per cent (Toyota 2004b, 39).

Much has been studied about this automobile giant. Toyota Production System is a symbol of efficient production, and its autos were reported, at least until 2010, to be more reliable than others. J.D. Power and Associates, a consumer rating company, ranked Toyota as the number one brand for passenger cars for five consecutive years. Additionally, 17 out of the top 38 models were from Toyota or Lexus (Toyota 2004b). This efficient, high quality, flexible production process, called "lean production," has been studied by more than a handful of scholars since the 1990s (Womack et al. 1991; Clark and Fujimoto 1991; Womack and Jones 1996; Cusumano and Nobeoka 1998; Fujimoto 1999). Two concepts of Toyota are widely known and transformed the production method of virtually every industry today: the *just-in-time system* and *kaizen*, or continuous improvement.

The *Just-in-time system* is essentially a set of practices and knowledge enabling a network of firms to produce and deliver products in small quantities and with short lead times (Liker 2004, 23). At every step of the manufacturing process, Toyota has a system that posts a signal (called *kanban* in Japanese) to the previous step that components need to be replenished. This creates a "pull" system in which production takes place only after consumption happens. Thus, Toyota accomplished its high productivity not by economies of scale, but by eliminating wastes and inventory management costs (Ono 1978).

The literal translation of *kaizen* is "improvement," yet it is more than an act of improvement, but a philosophy and set of practices intended to achieve continuous improvement. This philosophy differs sharply from the top-down management system for optimization based on Taylorism. According to Taylorism, a highly positivist paradigm that shaped the Fordist production in the U.S., management analyzes and establishes the most efficient system, and the factory workers are expected to implement it perfectly. In contrast, Toyota workers start with notions of problems. They assume no production system is perfect, but is full of problems. Finding, analyzing, and solving each problem is the daily practice of shop floor workers, and this continuous improvement would establish an efficient system (Ono 2001). This practice requires not only high motivation of every factory worker, but also the organizational structure to allow bottom-up decision making. Furthermore, it must be accompanied by daily practices in data collection, problem solving, and self-management within a small group.

For its strategy of internationalization, Toyota acted slowly and concentrated on the Japanese market in the beginning. Since the mid 1980s, Toyota has expanded international production, mainly in the U.S., and more aggressively plans to expand in non-Japanese markets over the next two decades.

Japan is still undoubtedly the largest market for Toyota (41.4 per cent of its sales), while North America is growing as its second major target, 34.2 per cent (See Table 4.1). The rest of sales are equally divided among Europe (11.7 per cent) and other areas, such as Asia (12.7 per cent). In contrast, this ratio for regional sales is different for U.S. automobile companies. Both GM (70.9 per cent) and Ford (56.4 per cent) rely more on their home market and sell more in Europe, but less in Asia-Pacific.

Table 4.1   Sales by major automobile firms by region

Firm	Japan	N. America	Europe	Other	Total
Toyota	41.4%	34.2%	11.7%	12.7%	100.0%
Nissan*	45.4%	32.3%	12.7%	9.6%	100.0%
Honda	23.0%	55.8%	9.3%	11.9%	100.0%
	Asia-Pacific	**N. America**	**Europe**	**Other**	**Total**
GM**	4.3%	70.9%	19.1%	5.7%	100.0%
Ford	4.8%	56.4%	36.8%	2.0%	100.0%

Source: Annual report of each firm.

Note: * Includes intra-firm sales.

** Sales by automobile division only, excluding revenue from other businesses, such as financing.

Toyota is now the largest non-U.S. car seller in North America. However, it has not been the most aggressive automobile manufacturer in the international market, particularly in the earlier period, compared to its rival, Honda. Honda experimentally launched production in North America in 1979 (Honda 2005), and quickly expanded its capacity in the 1980s and 1990s. As a result, Honda today sells more vehicles in the U.S. (55.8 per cent) than in Japan (23 per cent). Although Toyota started exporting to the U.S. in 1957, it was not until 1984 that Toyota launched production in the U.S. This started as a joint venture with GM, and they established New United Motor Manufacturing Inc. (NUMMI), in

Fremont, California. Accordingly, Toyota accelerated its international production by opening factories in Kentucky, and Indiana (see Table 4.2).

Table 4.2   Toyota production sites in North America

Name	Est. Year	Operation	Employment
Bodine Aluminum, Inc	1993	Aluminum castings	930
New United Motor Manufacturing	1984	Corolla, Tacoma	5630
TABC, Inc.	1972	Truck beds, catalytic converters	550
Toyota Motor Alabama	2003	Engines	340
Toyota Motor Kentucky	1988	Avalon, Camry, Solara, engines	7210
Toyota Motor Indiana	1998	Tundra, Sequoia, Sienna	4690
Toyota Motor Texas	2006	Tundra	2000
Toyota Motor W. Virginia	1998	Engines, transmissions	930

Source: Based on Toyota (2005b).

In the 1990s, Toyota shifted to a more aggressive strategy of international production under the leadership of a new president, Okuda. He aimed to "produce where demand is" and to balance the production and sales into three world regions: Japan, North America, and the others (Okuda 2004, 22–5). As a result, Toyota operated 42 factories in 24 countries by 2003, up from 11 production factories in nine countries in 1980. The next company president, Cho, followed this expansion strategy and announced a plan to double the current 2.6 million car production abroad by 2010 (Hino 2002, 297–8).

4.2 Toyota's R&D System

In 2004, Toyota conducted 54 R&D projects and spent $6.2 billion, the equivalent of 4.0 per cent of sales (Okuda 2004, 12).[1] Toyota operates these R&D activities at four levels: (1) a central research lab, (2) an advanced technical center, (3) four development centers, and (4) testing centers. It also owns a few overseas research centers; however, they are only for testing or exterior designing, not for product development.

---

1   This R&D ratio over sales appears to be lower than the industrial average of the automobile sector, 4.6%. However, this would not undermine the R&D intensity of Toyota because this is because of the vast scale of the firm in sales, US$166 billions. Second, this figure includes Toyota's production divisions, and seems to lower the percentage.

First, the Central Research and Development Laboratory conducts long-term basic research. Nine hundred and twenty-three researchers are located at Nagakute, 13 km northwest of the Toyota headquarters. They are involved in a wide range of automobile-related research, such as materials, air quality, mechanical, system, and electronics engineering (Toyota Central R&D Lab 2005). Research projects are commissioned by Toyota Motor or other group companies, such as Nippon Denso, a group affiliate specializing in electronics. The Advanced Technical Center of Higashi-Fuji, the second unit of Toyota's R&D arm described below, commissions 65–70 per cent of projects at this Central Lab. Approximately 80 per cent of interactions between Toyota and universities take place at this Central Lab (Harryson 1998, 151).

Second, Higashi-Fuji Advanced Technical Center hosts the Research and Development (RAD) group and specializes in the development of electronic systems and new materials. Additionally, this Center is in charge of bridging the basic research of the Central Lab and development centers, and its activities include advanced engineering development of specific body functions, such as body, chassis, engines and drive trains (Harryson 1998). Approximately 2,000 engineers are located at this site, about 200 km (125 miles) east of the company headquarters.

Visible product development takes place at the third level of the R&D arm, development centers, at the headquarters region of Toyoda. These centers are divided into four units according to platform design. Development Center 1 is responsible for rear-wheel vehicles, and Center 2 for front-wheel vehicles. Then, Center 3 works on sport utility vehicles. Each center has 1,500 to 1,900 engineers, working on about five new vehicle projects (Cusumano and Nobeoka 1998, 29). Previously, Toyota used to separate the divisions for designing bodies and interior equipment. This created coordination difficulties between the 16 different functional divisions. In 1993, they reorganized to the current structure by focusing on a product, and each division was limited to five functions: (1) planning, (2) body, (3) chassis, (4) power train, and (5) vehicle evaluation. For one car model, one chief engineer oversees the whole development process and is in charge of coordinating all engineering divisions. Figure 4.1 summarizes the relationship between engineering divisions and development centers.

Development Center 4 is the component and system development center. The establishment of Center 4 in 1993 simplified the work of the first three centers by separating the development of certain components and subsystems that could be achieved outside specific vehicle projects (Cusumano and Nobeoka 1998, 33). Thus, Toyota managers use Center 4 if they need a component not specific to a particular vehicle—a component that can be used in different vehicle models relatively easily. As a result, Center 4 engages more in development of components such as batteries, audio systems, and air conditioners, which do not need specific tailoring for each vehicle.

Lastly, Toyota operates vehicle testing centers—although they are officially called technical centers, little technical development takes place here. Shibetsu

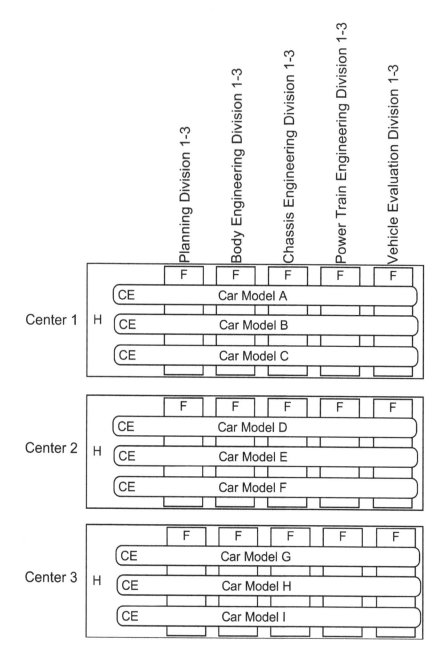

Figure 4.1   Product development organization of Toyota

Source: Modified from Toyota (1994).

Note: H-Center Head, CE-Chief Engineer, F-Functional Manager.

Center in Hokkaido, northern Japan, tests vehicles for cold conditions, while Whitman Center in Arizona does the same for extreme heat conditions. Toyota additionally organizes two overseas technical centers in Dearborn, Michigan, and Belgium. However, their roles are limited to testing components and materials, and researching emission standards, which differ in every country (Toyota 2004c, 8–9). Little basic research or product development is conducted outside Japan.

4.3 Overview of the Toyota Hybrid System

This section will briefly review the Toyota Hybrid System (THS). As the name suggests, there are two sources of energy: a gasoline engine (conventional), and a battery (new). Depending on driving conditions, the use of these sources changes. First, when a car starts, it uses the battery as an energy source. Second, as the car goes into normal cruising, it switches source to the gasoline engine. Third, when the car suddenly accelerates, it uses both energy sources. Finally, when it decelerates, it converts controlling energy into electricity, and recharges it back to the battery.

In sum, this was a system not only with a small, efficient gasoline engine and powerful battery, but integrating the two as one system. In addition to conventional combustion (engine), it required technologies in chemical engineering (battery), electrical, mechanical, and system engineering. It enabled the high efficiency of 51 miles per gallon, in contrast to 26 miles per gallon by the Toyota Corolla's conventional combustion engine.

4.4 Development of the Prius

Development of the hybrid car, the Prius, could be summarized as a story of complicated inter-divisional organizations. An automobile is a complex product with over 30,000 components, and Toyota mobilized more than ten development divisions to create major components and the hybrid system, as well as four factories. As organizational coordination took place frequently, through the trial and error of component and system development, all these units had to be located in the Toyota region. Additionally, the executives played a critical role in determining the features of the car in order to accelerate the whole development process, and to provide the financial resources needed for technological advancement, especially in simulation engineering.

4.4.1 Concept Development

This section will introduce two examples of a development process that required the co-location of project members and related divisions: a brainstorming exercise to create a concept image of the new car for the next century and inter-divisional coordination to design the system. To create the car concept, the project members proposed options, exchanged evaluations, and modified ideas. This was, in essence, an iterative process in which they had to be physically co-located and use face-to-face communication. To architect the system, Toyota established an independent unit from the conventional development centers. Nonetheless, this unit was located in conjunction with other development centers due to the need for frequent coordination. Therefore, both brainstorming and designing processes in this concept development required co-location of development units.

The origin of a car for the future using hybrid technology goes back to an executive order of the early 1990s. Interestingly, at this time, there was no concept of the Prius nor the slightest idea about hybrid technology. It was a former chairman, Eiji Toyota, who was concerned about the process and strategy for vehicle development at Toyota. He thought that the company, despite its steady business performance, was somewhat successful in developing and combining technologies, but only for marginal functions. Eiji worried that Toyota did not have a clear vision of what types of cars should be developed in the long term, and Toyota did not have the organizational flexibility to face this challenge. Thus, he thought there must be a new and fundamentally different concept for vehicle making for the 21st century (Ikari 1999, 26).

Toyota's executive board took this recommendation seriously and formed a commission, Global 21st Century (hereafter G21), in September 1993. Kanahara, the vice president who supervised the whole technology division, initiated this project group and appointed Kubochi, a director in the technology division, as the project leader. Kubochi was a veteran and respected manager who, for example, served as a chief engineer for the Celica and Carina, two highly successful models. G21 started with two layers of organization: The first consisted of senior managers from the technology management, body engineering, and engine technology divisions. The second consisted of junior level officers from body engineering divisions. Each group had ten members, and all were located at the headquarters.

At this stage, the organization of this commission was small, light, and flexible. It was a structure requiring a low level of commitment, and members of each subgroup met once a week near the headquarters. Members spent only part of their time on this project and continued their work in their originally affiliated division.

The goal of G21 was to find a concept for a car for the 21st century. While the group frequently and informally reported to the supervising executive, Kanahara, it produced one final presentation to ten executives in December 1993 (Itazaki 1999, 20). The concept of the new car presented could be summarized in two ideas: comfort and efficiency. It would be a comfortable car that people could get in and out of easily which means that the front seats should be relatively higher than usual. Additionally, the car for the next century must be fuel-efficient and to achieve such efficiency, it must be small; however, to provide comfort the interior had to be spacious. This was a somewhat contradictory idea, but the executives unanimously approved the overall concept. After this report, the group was dissolved, and each member went back to his original division. The first phase of G21 ended.

When the second phase of G21 began in January 1994, it required intense brainstorming meetings, which meant co-location of the project members. Moreover, it had two distinct organizational features compared to Toyota's standard product development process. First, Toyota decided to conduct product and concept development within this experimental set-up. Conventionally, product development takes place under a planning division at one of the three development centers (described in section 4.2 "Toyota's R&D System"). Since the new car would be front-wheel-drive, Development Center 2 would be a logical choice. However, Toyota executives were aware of limitations to this approach due to organizational constraints: Each established development center had its normal practices and interdivisional relationships, which could potentially limit creativity regarding the car for the new century. In order to pursue a fundamentally new car, Toyota formed a different organization to create concepts and coordinate development activities (Yaegashi, interview, December 4, 2004).

The second distinct organizational feature, related to the first point, was the appointment of a new project leader. The new leader, Takeshi Uchiyamada, was educated to college level in physics and was a 16-year veteran in the research and simulation of noise levels. Prior to the Prius project, he was assigned to the product evaluation and technology management division, and worked on organizational restructuring (Ikari 1999, 28–31). This appointment was unusual at Toyota because project leaders normally (1) had engineering backgrounds in engine, body or chassis design, considered mainstream in automobile development, and (2) had experience coordinating product development. Uchiyamada had little of either. However, the Toyota executives chose him because, wanting the new car to be original, they needed a project leader who would not simply follow previous practice and organizational ties within the product development divisions. Having a project leader with little experience in this area could mean fewer constraints on development, creativity, and core development organizations. Uchiyamada was an ideal candidate for this. In sum, the development team concentrating on G21 was a new, experimental, organizational attempt by Toyota to create a different type of a car.

At this time, the new team was more fixed in its make-up. Ten project members were assigned to G21 and while one member, Ogiso from body engineering, continued from the first phase, eight new faces joined. Almost all of them were technical engineers from various divisions of vehicle making. Their previous affiliations were the unit (component) production technology, body production technology, body engineering, body design, engine technology, drive train, and planning divisions. The mission of this second G21 was to draw up further concepts of a new car within six months. Therefore, project members were assigned full time, but expected to return to their original divisions after the appointment.

The second phase of G21 was also located in the headquarters district. An executive meeting room at the Third Technology Building was dedicated to this group (Yaegashi, interview, December 4, 2004)—a location which showed two features of the project. First, the group had higher priority than other development projects and second, the group had a separate room due to the confidentiality of the project. Conventionally, at the Development Centers, several development teams share an open space in a large room.

The first activity of the project requiring co-location was concept development. This was a brainstorming process in which project members gathered at one location to share ideas, feedback, and revise concepts. Compared to the original concept development phase, where members were only participating on a part-time basis, this second phase was more systematic, intensive, and comprehensive. The team started to list images of the 21st century, such as an aging society, variety of life values, increased roles of women, information technology, increased traffic accidents, and deteriorating traffic flows (Ikari 1999, 33). Then, the team conducted a series of brainstorm meetings to select the key concepts that the new car would answer to these images. Finally, what they came up was "natural resource and environmental problems." The rapid growth of the world's population and its increasing use of automobiles could exhaust petroleum reserves in the coming century. Global warming and the degradation of nature are increasingly important issues in both industrialized and developing countries. Thus, Uchiyamada lastly elaborated the concept as "a car that responds to environmental issues yet provides convenience and pleasure" (Yaegashi, interview, December 4, 2004). That would, at the least, mean a car that is minimally sufficient but not too small in size, with high fuel efficiency.

It was easier to achieve higher efficiency by reducing body weight and sacrificing speed and power. Uchiyamada estimated that 30 per cent higher efficiency (34 miles per gallon) than a popular medium sedan, e.g the Corolla (26 miles per gallon), would be possible. However, the team concluded that it would have only minimal impact in the marketing sense. If they wanted to claim a car for the 21st century friendly to the environment, 50 per cent improvement (39 miles per gallon) would be required. Since previously available engines could not meet this target, they chose a new D-4 engine, which would directly fuel gas inside cylinders, with an air to gas ratio of 50 to 1. This could achieve an extra 20 per cent efficiency.

At this stage of the concept development, the team members' activity was mostly on an individual level, and a formal all-member meeting of the group took place once a week. In the process of applying the whole car concept to a component level, the team set up goals, such as power levels, emissions volume, weight, and price. Since each member came from a specific engineering background, he was responsible for applying and estimating an assigned component. Thus, members collected and analyzed ideas from their original divisions and related research divisions. Here, knowledge was organizationally collected, but more precisely speaking, on an inter-organizational scale. The internal meetings of G21 were mainly to update one-another with each member's ideas. In the meantime, small meetings with two to three members took place whenever needed.

Intra-firm coordination was critical for making the Prius even at this early stage of the project, and this was the second process that required co-location of the divisions involved. As mentioned, each member of G21 provided component knowledge of each system based on his experience and affiliated division. However, since an automotive system is a complex system, the individual knowledge of the ten team members was not enough to configure the entire system. Therefore, they had to coordinate with and receive input from other key component divisions. In this coordinating work, Uchiyamada was a key person because his experience in the evaluation division had introduced him to engineers in various divisions. The team set up lectures by other divisions on seats, transmission, multimedia, and headlights, and thereby came up with better-configured system information on the car (Itazaki 1999, 46–7).

G21 was an early adopter of information technology—they quickly purchased two workstations for CAD. Kanehara, the G21 supervisor, did not hesitate over capital spending (Yaegashi, interview, December 4, 2004). In July 1994, G21 submitted a 200-page report of detailed concepts to the executives, which were accepted. Seven members went back to their original positions, and three stayed for the continuing work. The third and final phase of G21 started in August 1994, and finished at the end of 1995. Although only three members worked specifically for G21, a key decision was made for the creation of the new car: the use of a hybrid system.

By this time, Wada took over from Kanahara, the original board director responsible for G21, and became a board director of the technology division, as well as vice president. Uchiyamada frequently met with Wada to report on progress and exchange ideas. Most of the time, the two, who both had engineering backgrounds, agreed readily, but there was one issue on which they could not reach a consensus for a month, the targeted fuel efficiency. Wada believed that 50 per cent higher efficiency than the Corolla was not enough to advertise the Prius as a car for the next century and requested 100 per cent higher (22 km per liter or 52 miles per gallon). Calculating improvements in every possible component and the overall system, Uchiyamada could present only up to 50 per cent improved efficiency. This was rather an unusual situation. Normally, a chief engineer, with the spirit of an artist and challenge, would propose radical ideas, and an executive, as a manager, would scale them down due to considerations of time, cost, and feasibility—here, it was the opposite: Executive Wada proposed radical ideas, and Chief Engineer Uchiyamada had to advocate feasibility. When Wada threatened to terminate the project, Uchiyamada finally had to agree to proceed towards 100 per cent higher efficiency (Yaegashi, interview, December 4, 2004). As a result, they had to abandon the idea of using the D-4 engine, since it would be inadequate to meet this target, and had to look for alternative technology.

It was, in a way, surprising that the development team had not officially considered the use of a hybrid system up until this moment. However, Uchiyamada had reasons. The hybrid system, with a combination of batteries and gas engine, was still only at the research phase. He was aware that two teams within Toyota, the

Advanced Research Group at the Higashi Fuji Center, and the Electronic Vehicle Development Group, had worked on development of this technology, but they were still a long way from practical commercialization. Uchiyamada had to ask for an executive commitment to develop this technology and for the establishment of a new division to work on commercializing a hybrid engine.

There was an additional decision made by Wada which set the course of the project for the next two years. This was the decision to exhibit the car at the Tokyo Motor Show in October 1995 (Itazaki 1999, 54), at which major automobile manufacturers would present new car concepts for the next generation. It was at this time that the new car got its name, the Prius. The name Prius in Latin means "prior to" and Toyota's team wanted this car to initiate technology into the next century. Toyota strategically concealed all information relating to the hybrid technology at this exhibition, but advertised that they planned to introduce a highly fuel-efficient car by the end of 1997. The advertisement and commitment to this exhibition meant that G21 had to come up with at least a prototype in just one year, and swift development was necessary.

4.4.2 Search for Available Technology

Since the idea of installing the hybrid technology was imposed from the top, not from the engineers, the Prius followed a unique development process. First, there was a concept of the car, but without the technology—the search for and development of available technology happened after concept development. This occurred partly because the project was an experimental effort by Toyota. At the same time, it showed a serendipitous development pattern of product innovation. The nature of the product was not predetermined by existing technology, but the product concept called for new technology.

At this stage, there were two organizations that were formally working on the Prius: G21 and BR-VF. G21, in its second phase, had seven staff including four new full-time additions. Moreover, to develop the hybrid technology and system, a new group, BR-VF, was formed. This section will introduce two key features that contributed to product development. First, they needed to integrate electric, mechanical, and simulation engineering in order to establish the hybrid system, which required co-location of those members under BR-VF. Second, a close inter-divisional coordination between G21 and BR-VF was essential to advance the technology and set the course of the project. The organizational and financial support of the executives was key to facilitating the development process between these two units, as well.

In response to Uchiyamada's request, Wada established a new division to work on developing the hybrid technology, and it became a powerful partner to G21 throughout the process. Toyota had started organizational restructuring in the early 1990s. They created teams to analyze problems and propose alternatives, called Business Reform (BR) units. Toyota's technology division consisted of a total of 11 groups, and one team specifically focused on the improvement of the hybrid technology, BR-VF, with the latter letters standing for vehicle fuel economy (Matsui, interview, September 28, 2004). It started in February 1995 under the supervision of Kato, who was the second key executive in the development of the Prius. In order to establish the complex hybrid system, BR-VF needed to recruit staff from various engineering backgrounds. Three full-time members whose background was in engine systems, electric engineering for battery development, and mechanical engineering transferred from the Electronic Vehicle Development Division under Development Center 3 (Yaegashi, interview, December 4, 2004). Additionally, four electric and mechanical engineers assisted on a part-time basis, so the total staff came to ten as the new phase started.

Since there was no established hybrid system at Toyota or anywhere else in the world, BR-VF first had to analyze which hybrid system was commercially viable. They researched every possible hybrid system proposed around the world and initially found 80. Eventually, they narrowed these down to four systems based on mechanical theory and feasibility. Then, they had to generate a precise estimate of fuel use. Creating a prototype of each system would have been the ideal for the precise estimate and analysis of systems but this option was not feasible due to time constraints. They had to select and start development of a new system within three months. Therefore, they decided to use a computer-based simulation model via MatLab, software developed by MathWorks in Massachusetts.

At this stage, there were two types of inter-divisional coordination critical to the project, and both of them required immediate information exchange and decision making. First, close contact with the executives and technical divisions was essential to accelerate the simulation process. G21 requested the purchase of a new simulation software program and workstation. For such financial transactions, an executive body had to make a decision on such a purchase or an alternative quickly, and Executive Kato immediately approved a $300,000 purchase (Itazaki 1999, 67).

The connection between G21 and other technological divisions was also important for the development of this new technology. Sasaki, the electrical engineer in charge of the development of the hybrid system, started to use driving-simulation software, MatLab, for the project, and his first challenge was how to estimate fuel use under various driving conditions. There would be different scenarios for varying road conditions, such as an inner-city traffic jam, open space in a suburban area, highway, or mountain road. Other factors included weather conditions, the number of passengers, styles of driving etc. With endless permutations, he could not specify an algorithm program for the simulation. Sasaki's lucky day came when Arai, his former colleague at the Technical Center in Higashi Fuji who was researching other hybrid systems in Electronic Technology Division 3, visited G21 on the way back from the headquarters. Based on his past experiences, Arai presented a basic framework for an algorithm program with four major factors to determine driving conditions: (1) an accelerator, (2) a brake, (3) weight of the car, including the number of passengers, and (4) degree of slope in the road. With this framework Sasaki could set up a program with simplified yet reasonably effective simulation models (Muramatsu, interview, September 28,

2004). Here this exchange of information was random and came out of an informal network of friendship. However, it did take place as a result of a combination of the following factors: (1) the co-location of G21 and the company headquarters, and (2) a constant flow of people between the development divisions. Toyota provided an organizational platform which accommodated both factors.

There was another critical inter-divisional interaction between Uchiyamada of G21 and BR-VF. They organized meetings frequently, discussed the pros and cons of several systems, and finally selected a parallel hybrid system for the Prius. The physical distance between the two units was roughly 200 km (130 miles), connected by a major highway, but Toyota also operated a helicopter flight between the two locations several times a day. Therefore, it was a maximum two-

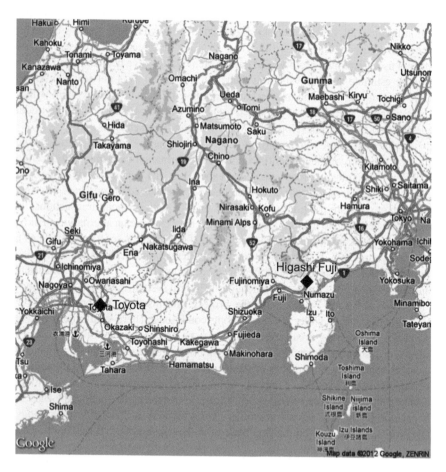

Figure 4.2   Map of Prius development 1

Source: Author.

hour journey with good transport connections. BR-VF presented a basic design and mechanics in June 1995—a high-level executive meeting at which all board members of technology divisions attended. Based on the feasibility of the hybrid system presented, they decided to upgrade the project from the research to the production phase and approved the necessary human and capital resources.

4.4.3 Component Development

The previous two sections described how two small groups, G21 and BR-VF, were formed and functioned. The main coordination style at that stage was the exchange of information between members within the same group as these groups conducted brainstorming exercises to develop product concepts and technology. Additionally, G21 and BR-VF coordinated with each other in order to integrate the core concept and technology. Moreover, a close relationship with the executive body was important to enable swift decisions regarding commodity purchases, and the setting of project and product goals. On the other hand, other inter-divisional coordination was based on personal contacts, such as between engineers who used to share the same office space. This level of coordination was effective in advancing technology at this stage, and organizational coordination was small in scale in the beginning.

Once the project went to the component development stage, aiming at mass production, the organizational coordination became far more complex. Since an automobile consists of 30,000 components, development of a completely new product unavoidably requires development of many new components. The number of divisions involved increased from the initial two to more than ten, and its coordination became complex. Inter-divisional coordination grew from the informal to a formal network of the whole organization: G21, BR-VF, engine development, suspension, drive train, unit production, production technology, electric vehicle development, body design teams, product evaluation division, and executives. For battery development, involvement by Matsushita Battery was critical, and the Calty Center in Los Angeles participated in designing the exterior body. Moreover, management of technological fields expanded to include electrical and mechanical engineering and chemistry.

From June to July 1995, Toyota organizationally assigned tasks to several divisions. As previously, BR-VF took the lead in developing the hybrid system. Other core technologies were assigned to several sub-divisions under Development Center 2 and Development Center 4.[2] The three divisions under Development Center 2 were Engine Technology Division 2—working on engine technology, Body Technology Division 2—working on brakes, and Body Evaluation Division 2—working on evaluation. The Electric Vehicle (EV) Development Division of Development Center 4 researched batteries and motors. Electronic Technology

---

2 See section 4.2, "Toyota's R&D System," for the organizational structure of Development Centers.

Division 2 and 4, and Hirose Factory co-developed inverters and semiconductors. Lastly, G21 continued to supervise the overall project (Yaegashi, interview, December 4, 2004).

From July 1995, BR-VF went into a second stage as it shifted from research to development. BR-VF did not have the resources to develop physical components, but Toyota coordinated by transferring a total of five members out of BR-VF; each member was to develop specific components.[3] Four members who were affiliated to BR-VF on a part-time basis went back to their original divisions: the Unit Production Technology Division and the Production Technology Development Division of Development Center 2. Additionally, three new staff from the EV Development Division came to BR-VF to create the first Prius prototype, to be tested in November. These three were in charge of developing software for the electrical system, system control, and braking (Ikari 1999, 51). At the same time, Matsui, who was affiliated to BR-VF full time, was assigned to the Drive Train Technology Division to establish a planetary gear mechanism, a key mechanism in hybrid technology, by integrating the engine, generator, and motor.

In addition to the human resource transfers, the coordination between BR-VF and EV Development took place through dense and frequent communication. Matsui of BR-VF described it as "a catch-ball process" with Kubo, his counterpart at EV Development (Matsui, interview, September 28, 2004). As an example, a redesigning sequence of an engine system would exemplify this playing catch process of these two divisions. An engine system is composed of an engine, transaxle, damper, generator, planetary gears, decelerator and motor. Conventionally, all these components would line up adjacent to each other. However, they had to scale down the engine system so that it would fit into the small front engine space of the Prius. They redesigned the size and shape of each component a number of times, and tested to see whether it would function efficiently. After many trials, they decided to relocate the decelerator, outside the engine compartment, but connected with 20 mm chains.

In December 1995, Toyota's top executives made an important decision to accelerate the project by releasing the Prius by the end of 1997. While Toyota had been known more for its product reliability than for product innovation, both Toyoda Shoichiro, a chairman of Toyota, and Hiroshi Okuda, a newly appointed CEO in August 1995, showed a strong interest in getting the hybrid car on the market ahead of any. They pushed the release date to more than a year earlier than any project members had expected. Organizationally, this decision led to three changes. First, it shortened the project period so that Toyota as a whole must conduct component development and preparation for mass production simultaneously. This simultaneous engineering will be further discussed in the following section. Second, with their commitment and support, Toyota executives made the unusual announcement that the Prius development would be a high

---

3 Earlier, Toyota attempted to promote interaction between BR-VF and EV Development Division by having the same director, Fujii, in charge of two divisions.

priority project for the company. They guaranteed any support in terms of financial and human resources, and called for the swift cooperation of all divisions. Third, they quickly upgraded the two core groups within the project: G21 and BR-VF. In January 1996, G21 was renamed Zi and structured under Development Center 2, a division for developing front-wheel-drive vehicles. This meant that the whole project was now preparing for the production line. The quasi-experimental and developmental phase of the project ended, and the official, full-scale mobilization started. Additionally, Zi recruited a new member, Oi, who was a specialist in coordinating various component-developing divisions, factories, and suppliers (Yaegashi, interview, December 4, 2004).

Similarly in February 1996, BR-VF was upgraded and welcomed a new leader, Yaegashi, a veteran with extensive experience in emission control for engines. He reluctantly accepted the appointment, as he considered there was only a 5 per cent chance of completing the Prius by the end of 1997 (Yaegashi, interview, December 4, 2004). The rush project was considered unfeasible even among Toyota's own engineers. Nonetheless, the Toyota executives promised their full support and accepted ideas that Yaegashi proposed. Essentially, he proposed the creation of three new groups to support BR-VF and its coordination with other divisions. The addition of more divisions would make organizational coordination even more complex, but it was unavoidable in developing a technologically challenging system. The first group was in charge of managing the reliability of the Prius, in cooperation with Engine Technology Division 1. The second group aimed at coordinating the hybrid system (Toyota Hybrid System, or THS) with other components in the car. Since this coordination went beyond engineering for engine technology, they integrated control experts, located in Higashi Fuji. This team used software programs to simulate functioning and safety. The third group was designated to look for any potential deficiencies that the Prius might have. By this time, the full-time staff in BR-VF had increased to 13 engineers. The backgrounds of newly recruited engineers ranged from engine and braking technology to electronics and production technology. In total, 50 people, including part-time workers, got involved in designing the THS (Muramatsu, interview, September 28, 2004).

Geographical proximity was critical for these divisions working on integrating the engine and THS. While G21 was located near the headquarters, other groups were located in Higashi Fuji and coordinated from a distance at the beginning. Soon, this two-hour distance became a burden for project members as they had to integrate not only the engine system, but also the motor, battery, and gear mechanics. Thus, in January 1996, groups from the Higashi Fuji center relocated to the EV Development Division near the company headquarters (Yaegashi, interview, December 4, 2004).

The Prius test runs show how product development started prematurely and evolved through stages. The first Prius tested in January 1996 was very different from when it was commercialized in October 1997. Since Toyota did not have sufficiently powerful batteries at that time, they had to run a series of batteries in parallel. The volume of these was enormous, and the entire trunk space was dedicated to housing them. Moreover, the computer program for controlling the engine system was under development, and they had to use five or six desktop computers which occupied the whole back seat (Muramatsu, interview, September 28, 2004). At the testing center in Higashi Fuji, this heavily weighed-down car of the new century experienced its first *movement*. There were many problems, and it was hardly a smooth or fast run. Nonetheless, the project members all celebrated the first run of a new car.

The creation of a new battery for the Prius was an immense challenge to Toyota and led to intense interactions between Toyota and Matsushita Battery, a subsidiary of Matsushita Electronics, a consumer electronics giant. This inter-firm coordination was critical to developing the batteries—a field in which Toyota had not traditionally had strength. Conventional cars use batteries which produce only 12 or 24 volts, which would be needed only to start the car or to provide lights for a short time. Since the Prius was designed to run without a gasoline engine in some situations,[4] its batteries needed to produce 288 volts, approximately 25 times more. In its TownAce EV of 1993, Toyota did have a vehicle which ran on battery electricity only. However, this electric vehicle had several limitations. First, as the capacity of each battery was limited, a number of batteries had to be stored to produce sufficient electricity. The vast volume and weight of the batteries severely limited the mileage of the car per battery charge. Additionally, recharging batteries took hours and was not convenient at all. Toyota, Honda, GM, and Ford all announced electric vehicles in the 1990s. However, these were more for advertisement purposes, to be benign to the environment, than to produce on a mass scale and make a profit. Toyota saw the Prius differently. It intended to commercialize this brand in the mass market from the beginning, and therefore had to produce a smaller and more powerful battery unit.

The first meeting between Toyota and Matsushita took place in March 1995. Toyota asked whether Matsushita Battery could produce batteries with the following demanding specifications: (1) that it be smaller, about half the size of what was previously available, (2) made of nickel metal-hydride, which is more environmentally friendly than nickel cadmium, (3) twice as powerful, (4) durable between —30°C (winter) and 60°C (summer with engine heat), and (5) reliably rechargeable. The Research Lab at Matsushita Battery, located in Moriguchi, Osaka, 231 km (144 miles) west of Toyota (2.5-hour journey), started this challenging project with a few staff in September 1995. Each time they developed a new prototype, they carried it from Osaka to Higashi Fuji for test runs. However, they could "not achieve required power or reliability in the first year, and it was a tough time" (Ikoma, interview, September 28, 2004). Both Toyota and Matsushita staff visited each other frequently, discussing weaknesses and possible improvements, sometimes all night. At its peak, more than 50 engineers were involved in the battery project, and Matsushita Battery proposed 15 prototypes.

---

4   See section 4.3 for more details of the hybrid system.

The battle over batteries continued for about two years. Toyota was an innovator in simultaneous engineering project management (Ward et al. 1995) or concurrent engineering (Sobek et al. 1999). They developed components with two possible designs, i.e. with a plan B in case plan A did not work. The Prius team prepared for an ideal scenario (Plan A) in which batteries of the expected capacity would be produced, as well as for the worst-case scenario (Plan B) in which the power of the battery would be only half what was needed, meaning that they would have to store twice the volume of batteries. Consequently, they prepared two designs of the engine system depending on where the batteries would be located. By late 1996, the efforts of project members became fruitful, and they produced batteries with the required power. Though this was a big step to clear a tough hurdle, it was only the mid-point. Next, Toyota had to develop a system to control the batteries.

Inter-connectivity of various components for a new product would be best described by the story of the battery and its heat problems. The heat level and voltage of batteries increase exponentially when they are almost fully recharged. Therefore, the battery group had to develop a method, first, to precisely measure the state of charge, and second, to control the charge level. It was an even greater technical challenge because the heat increase and charge level of nickel metal-hydride batteries would depend on the *outside* temperature. Additionally, the battery team had to coordinate with a group specifically dealing with the cooling system in order to work out the most desirable location for the batteries. Originally, they intended the batteries to be stored under the front seats, as Toyota did with its previous electric vehicles, such as RAV4-EV. However, using this space was troublesome because the air coming out of the front space would be warmed up by the heat of the engine. Wada, a board member, intervened and advised that the batteries be located between the back seats and the trunk (Yaegashi, interview, December 4, 2004). That space was more appropriate because it would not be affected by the heat of the road surface and it could easily receive air-conditioned air. This battery space issue came to an end only after confirmation by several experiments and design changes in the engine system. Adding the cooling system changed the size and weight of the battery system. The weight of the batteries alone was 45 kg, but the system ended up weighing 75 kg (Uchiyamada, cited in Toyota 2000, 91).

Innovation may grow out of unintended outocmes. The engine-development team unexpectedly developed methods to control the batteries from their testing experiences. Nickel batteries are known for their memory effect in which the life of batteries decreases as used. After a number of experiments, the engine team discovered a new phenomenon which could control the memory effect. If the charging level of batteries was kept in the middle, the memory effect could be minimized. In other words, it would increase the durability of batteries. Therefore, a system to control the battery charge level was critical, and BR-VF mobilized resources for its development.

As both Toyota and Matsushita Battery were becoming confident about the quality of the batteries, they launched a joint venture.[5] This new firm, Panasonic EV Energy, was established in December 1996 and located at Kosai, Shizuoka, 56 km (35 miles) southeast of Toyota. Fujii, the director of the EV Development Division, primarily led this newly created company (Panasonic EV Energy 2005).

The development of the Prius is the story of how a number of development team members benefited from co-location. In this sense, geographical concentration was the key to promoting coordination. But there was one exception to the co-location rule—its exterior body design was developed overseas. The major ideas for the final design came from the Calty Design Center in California through Toyota's internal bid process. Later, a design team under Development Center 2 collaborated and finalized it.

Toyota had seven design centers in the world: one under the Technology Division, three under each of Development Centers 1 through 3 (all these four located in the Toyota region), Tokyo Design Center, Calty Design Center in Torrance, California, and Epoch Center in Brussels, Belgium. Initially, Uchiyamada assigned the task to the group under Development Center 2 in December 1995 (Yaegashi, interview, December 4, 2004). This was a logical choice and an organizational fit because all front-wheel vehicles, including the Prius, were developed at Development Center 2. As stated in the section of "Concept Development," the Prius had to meet the following four concepts: comfortable, convenient, small in overall size, but sufficiently spacious inside. It was expected that the new car would be somewhat shorter than the Corolla, but as spacious as the Camry. The three-staff team submitted several designs in early 1996; however, none of them satisfied the chief engineer. Uchiyamada considered that their designs did not fully present a new, futuristic style or appealing appearance. He decided to open up a design bid within the Toyota Group, which had three rounds of reviews.

In the first round of competition, February 1996, seven teams submitted models at 20 per cent scale: the Headquarters Technology Division, Development Center 2, Tokyo Center, Calty, Epoch, a team of contract-designers, and interestingly a design team outside of Toyota Motors yet under the Toyota Group, Toyota Automatic Loom Works. Designs from four teams passed this round, and they submitted clay models at 100 per cent scale in May for the second round. Two plans received equally high evaluation: one by the Headquarters Technology Division, which emphasized friendliness and familiarity, and the other by the Calty Design Center, which presented a unique, futuristic style. The third round, the most important and comprehensive evaluation, was in July 1996, and involved several key board members and 200 panel reviewers from various sections within Toyota. They assessed several dozen aspects and gave points on a scale regarding various factors such as innovativeness, future orientation, and appeal to youth and women. Both ideas received high scores; however, the design by Irwin Lui of

---

5   Matsushita Battery invested 36 per cent of the capital, Matsushita Electric 24 per cent, and Toyota 40 per cent. (see in Itazaki 1997, 200).

Calty Design was finally accepted because of its radically new style, symbolizing the challenging spirit of Toyota toward the new century (Ikari 1999, 138). This became the basic exterior design of the final Prius running on the street.

The exterior design was drawn almost independently. G21 (more recently, Zi) provided the basic information on the car at the beginning: concepts and overall exterior scale. Based on statistics and ergonomics, they designed to fit from a small Japanese woman (148 cm or 4'10") to a tall American man (190 cm or 6'3") (Hoshino, interview, April 24, 2004). Then, each design group, including Calty Design, regenerated concepts into design, proceeding on their own. Thus, the interaction between them was minimal here. The Calty Design Center had a total of 50 designers, and about half of them were Japanese staff. For this project, Irwin Lui, the chief designer, and a group of five people participated in the design process (Hoshino, interview, April 24, 2004). Once the Calty design was selected, there was some coordination between Lui of California and Takekuni Saito of Design Division 2 under Development Center 2. The original exterior design had to be modified to make it suitable for mass production, and a team led by Saito conducted this detailed redesign. Long-distance communication using Email and graphics files was sufficient for this process, and no face-to-face communication was involved. It was only after two months that Irwin Lui travelled to Toyota City and saw the final design in September. Irwin was satisfied that Saito's team understood his original idea and elaborated (Hoshino, interview, April 24, 2004).

One last major unit involved at this stage was the production evaluation group. This group did not develop or produce any components. However, it was another key unit that all component development groups had to coordinate with. All components needed to pass the evaluation conducted by this unit. Through testing, they would identify problematic components and call for improvements. This identification process was crucial because, for the development of a car with an entirely new system such as the Prius, there were a number of technological problems at the testing phase. The causes of these problems were not easily identifiable as they could be caused by individual or multiple component malfunction or lack of coordination between components. There were seven staff members at the product evaluation unit under Development Center 2: Two of them worked full time on the Prius, with the rest involved a fair amount of time.

Toyota created product evaluation groups in the early 1990s for two reasons. Firstly, to evaluate each component and the overall integrated system from the viewpoint of consumers, not that of engineers. Quite often, well-engineered components might not be the most convenient or comfortable for consumers. It was Toyota's intention to incorporate the consumer's perspective even at the product development stage. Secondly, the evaluation unit was intended to help chief engineers by providing recommendations. As the automobile components and system became more complex, so did the technology and organizations within the company. Different divisions had different objectives and might disagree on complex issues: One might insist on a specific technology convenient for their own purposes but neglecting the overall benefits and costs for a larger system.

For example, the suspension (between the wheels and the chassis) needed anti-vibrating rubber and for safety reasons, and this rubber needed to be less elastic, while more elastic rubber was preferred for handle-steering matters. Therefore, it would often be difficult for engineers in the safety regulation and drive-train divisions to reach an agreement, and a chief engineer had to intervene and make a final decision. It was here that the product evaluation units could help the chief engineer by researching, analyzing, and making recommendations. As a result, this evaluation unit needed to know about all sorts of matterss: about various components, technologies, and the people within many divisions. Uchiyamada, who had been in this unit from 1992 to 1997, recalled that his experiences there were invaluable to his work as a chief engineer, since he knew key people who knew about miscellaneous component and technology issues (Yaegashi, interview, December 4, 2004).

Discussing the Prius engine's shock problem and its solution will show why this evaluation unit and the testing they carried out was invaluable and how the different development units worked together. By this time, the fundamentals of the hybrid system were established, and the Prius could run with both a gasoline engine and a battery-powered motor. Therefore, the system physically worked from the perspective of each power-generating component. However, the evaluation team soon found that the driver faced a physical shock, like stalling, when the gasoline engine was turned off, and the car switched to battery power when operating at low speeds or going downhill. Such a shock would be both uncomfortable and not desirable from a consumer's point of view. The engine and battery divisions and BR-VF formed working groups to come up with a solution.

After some weeks of trials, they found that reducing the airflow into the engine brought less oxygen and lower air pressure would cause less ignition and decrease physical shock. First, they looked at the angle of an air intake valve—by changing the cam angle in each test, they found that the shock was reduced. Second, they came up with an idea to integrate the hybrid system with an adjustable air intake valve—this valve allowed only a small amount of air initially into the engine compartment, but increased it gradually. One of the working groups was aware that this adjustable air intake technology was called Variable Valve Timing organization (VVTi) and was used for the Mark II and Crown, two high-class cars manufactured by Toyota. After a year of trials, the groups eventually incorporated these two elements so that there would be minimal shock when turning the engine off or on when driving (Kotani, interview, September 29, 2004). Producing a component was not sufficient. Producing a prototype as a complete system and testing it revealed problems and the actions necessary for improvement. In total, Toyota produced 60 prototype cars and hundreds of engines (Uchiyamada, cited in Toyota 2000, 90).

4.4.4 Toward Mass Production

In addition to the ten divisions already involved, this mass production phase required the participation of the final critical units, the factories. This section will first introduce two organizational examples from Toyota that show smooth organizational coordination between development units and factories: the Unit Production Technology Division and the resident engineers. Unlike the Vaio case, in which Sony carried out production in a distant location, Toyota integrated development and production by co-locating them. To demonstrate last-minute changes in design production, the section will discuss problems found in the cooling system and the use of computer-aided design. These changes required coordination and readjustment of all the divisions involved, which they could only operate efficiently when in proximity.

As previously stated, Toyota has been the world's innovator in simultaneous engineering (SE). This is an engineering and management attempt to develop and produce components at the same time. In this sense, it is hard to draw a time boundary between "component development," "bridging to mass production," and "mass production." Technically, it was soon after January 1996, when G21 was upgraded to Zi, that Toyota started to prepare for mass production. The critical difference was that during the component development phase, each division produced components on a test basis, and the number of components in production was fairly small. On the other hand, during the preparation for mass production, each production division led by G21 aimed to prepare for producing 1,000 cars per month.

To promote simultaneous engineering, Toyota had two distinct organizational systems: production bridging organizations and resident engineers. Like many manufacturing firms, Toyota traditionally had a production technology division, which researched and introduced methods for production at their factories. While this division could speed up the launch of factory production lines, Toyota considered it insufficient. To further enhance the link between the development divisions and the Production Technology Division, they established the Unit Production Technology Development Division and Body Production Technology Division in the 1990s. For the Prius project, Tanoue, an engine development engineer, transferred from Development Center 2 to the Unit Production Technology Development Division. He stated that the "most important objective of simultaneous engineering was to advise development divisions from the perspective of the production side from an early stage" (Tanoue, cited in Ikari 1999, 77). Development engineers were, after all, development-oriented. In other words, their primary concern was the function of a component and not its availability or whether that component could be mass-produced practically. A few simple suggestions from the production side at an early stage could alter the designs of components, which was far more efficient in time and cost than altering designs later.

The second organizational system to speed up bridging to mass production was the use of resident engineers (REs). These REs were technology engineers from the development divisions, but they would work at the factories to advise on the alteration of design or production lines. In some cases, resident engineers at a single factory could make more than 1,000 suggestions in a given month (Hino 2002). In addition to resident engineers, the Prius project used "counter resident engineers." This meant that production engineers from the factories would come to the development divisions to further accelerate the bridging process by familiarizing themselves with components and technologies, as well as making suggestions. Mayumi and Matsumoto of the Quality Management Division at the Takaoka Factory were voluntarily stationed at the Zi office.

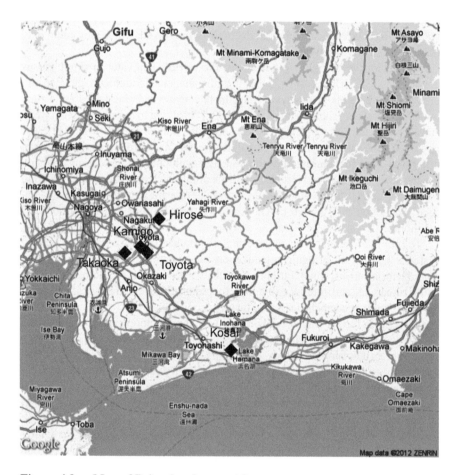

Figure 4.3   Map of Prius development 2

Source: Author.

The coordination organized under G21 reached new levels. While previously it had been restricted only to development units, it was then extended to factories, sales divisions, and even suppliers. To deal with this extended, complicated coordination, G21 received new members in January 1996. First, Oi, a specialist in coordination between the development and production divisions. In the past, he had been assigned to launch manufacturing of the Tercel in 1994 and the Starlet in 1995. These were two small popular cars in Japan and had many structural similarities with the Prius. Ironically, as the Prius project was conducted clandestinely before 1996, Oi knew nothing about a hybrid car or the Prius. In December 1995, he received a phone call from his supervisor and started to look for a secret office of Zi. His first job was to understand what the Prius was. Two additional team members were counter resident engineers from the Takaoka Factory, coming under the direct supervision of Oi.

Of the 12 factories in its headquarters region, Toyota mobilized four for mass production of the Prius. The Kamigo Factory worked on engines and the Hirose Factory on inverters and Electric Control Units (ECUs). The Headquarters Factory, normally producing chassis for commercial vehicles, was specifically assigned to produce motors and transaxles. Considering the hybrid car as a vehicle for the coming century, Toyota wanted its historic Headquarters Factory to produce one of the major components. Additionally, this factory had the advantage of geographical proximity to other key divisions: Development Center 2, the Production Technology Division, EV Development, BR-VF, and Zi. Since a motor is a key component with a number of new technologies, proximity was important in order to call up component designers whenever trouble happened (Toyota 2004c). Lastly, the Takaoka Factory would function as a final assembly plant. All four factories were located at Toyota City.

Table 4.3   Factories to produce the Prius

Factory Name	Main Products	Year Established	Site Area	Employees*
Headquarters	Vehicles, chassis, forged parts	1938	550	2,598
Kamigo	Engines	1965	870	3,177
Takaoka	Vehicles assembly	1966	1,360	5,370
Hirose	Electronic parts, semiconductors	1989	250	1,299

Source: Toyota (2005c).

Note: * This is the total number of employees per factory. Only some of those worked on Prius.

The Takaoka Factory started preparing for assembly in June 1996 (Kaneda, interview, September 29, 2004). This factory was selected for the final assembly, first, because it was well known for its high quality production, and second because it manufactured three popular small cars, the Corolla, Starlet, and Tercel, which were on a similar scale to the Prius. However, the Takaoka Factory had to face a major rearrangement of its production line. Toyota's assembly factories normally used mix-lines, in which each line produced multiple types of cars if they used the same components. This was a strategy to save factory space and give more flexibility in production volume. On the other hand, the mix-lines between the Prius and other models would not work because of the big difference in production volume. The factory had capacity for assembling a maximum of 1,000 cars per day for other models, while the expected volume of the Prius was only 1,000 per month. In order to set up a separate assembly line for the Prius, managers from the production of body, press, welding, painting, rigging, and testing teams, as well as the development engineers for related major components held a meeting in July 1996 (Kaneda, interview, September 29, 2004). They discussed detailed plans for assembly, such as which order each component would come in, where and how much space each process would occupy, at what height and position each worker in each process would operate, etc. This week-long meeting proved to be intense and required commitment from all the project managers involved. All key parties needed to be present and to exchange opinions, long-distance communication was not an option; face-to-face communication was irreplaceable. Additionally, it required organizational flexibility from all the divisions involved so that they could attend whenever a brainstorm meeting was called for. Furthermore, it was not only at this early stage, but also for the continuing process that the setup of production lines required coordination between the production and development divisions. Since components were still being developed and redesigned in the meantime, there was also the need for the readjustment of production lines.

A change in design and component location could alter the whole assembly line, and continuing coordination between production and development divisions was needed. The case of battery location is an ideal example of this connection. As previously stated, batteries were designed originally to be under the front seat, but moved to behind the rear seat because of connections with the cooling system. That decision caused major changes in the setup of assembly lines when it came to production. Kaneda at the Takaoka Factory stated that:

> That [battery relocation] was a major problem for us. We first prepared the assembly facility for the bottom of the car, but it was switched to behind the rear seat. We had to rearrange most assembly facilities in terms of components, order, and location. Along with that, we had to rearrange body welding facilities. If it were the bottom of the seats, that would have been easy coordination between the bottom plate and batteries. Now switched to behind the rear seat, we had to redesign with the bottom plate, internal body, and batteries. This additionally

required further readjustment of the facility in relation to the locations of supporting components of batteries (Kaneda, interview, September 29, 2004).

Changing assembly line facilities meant changing jobs for related component developers. All managers of the Takaoka Factory, production technology, and battery divisions had to be physically present at the site to discuss and confirm the finished setup of the assembly lines. Another reason for Toyota to select the Takaoka Factory for its proximity to other divisions involved.

There were problems rooted in component development, but they could identify these and adjust to them only after production had started. Two examples will be considered: (1) issues with a cooling system, and (2) mistakes related to the use of computer-aided design (CAD). These examples show that a tight linkage between development divisions, as well as between development divisions and production lines, was key to product development and transition from development to mass production.

First, the issues with the cooling system were a classic example of a gap between designing a single component and integrating several components into a system. An automobile requires a cooling system, traditionally for both the engine and passenger space. The Prius had several extra components that required cooling: a motor, inverter, and batteries. The cooling system for the batteries was important because a high temperature would lower the recharging level of the batteries. Toyota formed a new group to deal with the cooling system: six specialists from the batteries, motor, inverter, and engine divisions (Ikoma, interview, September 28, 2004). They decided that the batteries would be cooled with the air-conditioned air from the passenger space. Therefore, they had to install a system to monitor both the passenger space and batteries, and balance the battery use, AC level, and outer temperature. They additionally had to establish another cooling system for the inverter and motor. They designed two cooling systems based on the function, energy consumption, and temperature increase expected for each component. Up to this point, things went as planned.

However, when they made a second prototype in June 1996, they realized that the temperature was 5°C higher than it should have been. There were two unexpected reasons for this. First, as each component was improved day by day, its capability got better, which increased the necessary energy consumption and thereby the temperature. Second, when they tested those components and the cooling system in the first prototype in late 1995, it was during Winter when the outside temperature was low. June is Summer in Japan, and the cooling system hit its limit sooner than originally designed (Muramatsu, interview, September 28, 2004). They had to redesign and upgrade the entire cooling system so that it could function under any weather conditions whatever the season.

This case with the cooling system showed that just designing and estimating the workings of each component had limitations. They needed to, at least, build a prototype so that components could be seen when functioning within the whole system. Moreover, other unexpected aspects, such as improved components and a change in seasons, could alter the estimates. They could adjust to these ongoing changes only after component development was completed to some degree as the original plan often did not work. The development and production teams had to be flexible in their approach to development.

The use of computer-aided design (CAD) illustrates how critically information technology can be tied to product development, how much it can improve productivity, yet how it sometimes cannot solve every problem. The introduction of CAD dramatically improved the redesigning process. However, it did not alter the importance of geographical proximity and created another problem. Compared to their rivals in the industry, Toyota was slow to implement information technology for product development and only introduced it in the early 1990s (Cusumano and Nobeoka 1998). This lack of interest in more sophisticated computer technology was mostly due to their core strength being based upon a finely tuned development process with well-trained engineers. They emphasized the organizational practice of reviewing design and finding mistakes and rather superficially added information technology to enhance their development and production process (Nobeoka and Fujimoto 2004, 16–17). Nonetheless, they launched a major virtual designing system, Visual and Virtual Communication (V-Comm) in 1996. The purpose of this system was to conduct design, prototyping, testing, and production trials virtually in order to reduce the number of prototyping and design changes. The result was substantial especially for the engine compartment, which was a component-dense space, as well as for the exterior body, which required precise designing between parts, such as between doors. After switching to the CAD system, the number of redesigns was reduced from 10,000 to 500 per model (Hino 2002, 314).

On the other hand, the use of CAD created a major problem in the Prius case. In December 1996, Matsui at the Drive Train Technology Division received a phone call from the Headquarters Factory, which was testing production and faced a problem. The problem occurred because one of the planetary gears and its outer container collided, and would not physically fit together. It was a simple design mistake resulting from using CAD. For gear design, it was common to draw from one surface by using two-dimensional CAD and rotate 360 degrees, assuming that the same design applies to all four surfaces of a component. However, for the planetary gears on the Prius, one of the surfaces had to be redesigned. One corner space had to be curved and reduced differently from the others, and the gear designer had overlooked this. Although this mistake was elementary, technical, and avoidable, the financial, time, and organizational cost was significant. To fix it, Toyota required organizational mobility and swift coordination among several groups.

Matsui and another veteran designer rushed to the Headquarters Factory and called an emergency meeting among the divisions involved: the Drive Train Technology, Headquarters Factory, Unit Production Technology, and Akashi Factory. The Akashi Factory, producing casting metals for the gears, was located in Miyoshi Town, next to Toyota City, and still within the immediate vicinity of the headquarters. To change the production of the outer case, they had to change the casting metal. However, this change of casting metal would take one to two months, and they decided to use the new casting metal only for mass production, but not for testing models. In the meantime, they decided to fix the shape of the outer case by hand and use these gears for testing in January 1997. Several staff from the Unit Production Technology and Drive Train Technology Division came to the factory on the weekend to work on this repairing operation (Kaneda, interview, September 29, 2004). Here, organizations had to have flexibility to instantaneously mobilize staff and to solve problems, often in the format of a face-to-face meeting. To prepare for this kind of emergency, firms benefit most by locating all the potentially involved divisions in geographical proximity.

There were more problems found even closer to the market release of the Prius. The evaluation group was key to finding problems and making improvements as the product release approached. Additionally, some problems could require assembly adjustments that would be too costly at the factory level. In that case, this readjustment took place outside Toyota, at its suppliers, in other words, before assembly took place. Needless to say, this required more extensive coordination between Toyota and its suppliers. For example, the evaluation team found a sound problem with relays. These relays controlled electricity inside the car system— between different voltages and preventing sparks. Almost all cars used three to four relays in each car, yet the sound of turning the relays on and off was never a problem because the louder engine would conceal the relay sound. In this case, this relay sound was quite noticeable since the Prius would sometimes run without a gas engine, using batteries only and was therefore much quieter. This problem was foreseen by neither BR-VF, the system integration group, nor Zi at the system planning level; only prototype making and testing revealed it.

The problem was reported to Oi of Zi, and Uchiyamada directed engineers to soften this sound problem. Their first solution was to cover the relays with sound-absorbing insulators. However, the Takaoka Factory's assembly lines were fixed by this time, and adding another process was not optimal (Yaegashi, interview, December 4, 2004). Then they decided to fix the problem before the component arrived at Toyota. They asked Anden,[6] a component supplier of relays, to use sound insulator. This case shows that, first, Toyota and its suppliers had to have a close, coordinated relationship, and second, both sides had to be organizationally flexible to cope with unexpected incidents. The noise problem of the relays was technical, but to solve this problem, Toyota had to come up not only with a technical solution (i.e. using sound insulators), but also with organizational adjustments (i.e., requesting a further process from its supplier).

These last-minute production problems were not uncommon. The chief engineer, Uchiyamada, and Oi, who was in charge of overall component procurement and supplier relationships, stayed at the Takaoka Factory every day for the final month before the Prius debut (Uchiyamada, cited in Toyota 2000, 93).

---

6 Anden is a 100 per cent subsidiary of Denso, Toyota's major electronics supplier.

The Prius finally came on the market on December 10, 1997. Toyota President Okuda proudly presented the car for the next century as "[environmentally] harmonious growth" (Okuda, cited in Toyota 1997), and it overwhelmingly received support for the Japan Car of the Year Prize in 1997 (JCOTY 1997), as well as the New Car of the Year Prize by the Automotive Researchers' and Journalists' Conference of Japan (RJC 1998). It was the first car in history to steal both prestigious prizes. At the same time, the Prius was presented in U.S. automotive shows, such as the Electric Vehicle Symposium in Orlando, Florida, and received the Climate Protection Award from the U.S. Environmental and Protection Agency (EPA 1998).

4.5 Summary

In its over-three-year effort to combine conventional combustion and new battery technology, Toyota mobilized more than 300 development-related engineers and an additional 700 production-related engineers (Ikari 1999, 195; Itazaki 1999, 16). Geographically, Toyota concentrated most development functions in Toyota City, including G21 (later Zi) and all units under Development Center 2, as well as all its production capabilities in four factories. There were major research units located at Higashi Fuji, such as testing and early technology development. Yet this Advanced Technical Center was still located within a two-hour distance, with major highway access and a constant flow of helicopters. On the other hand, their new core technology development unit, BR-VF, moved to Toyota City by 1996. Even the two-hour distance was becoming a major obstacle for closer interaction with other development units, and co-location was their solution. Only the exterior design was conducted outside the Toyota core region. After a series of internal competitions, Toyota chose the design by the Calty Design Center in California. Although the exterior design was highly visible and important to consumers, this process was considered rather minor and conducted independently from other technological development, after the main development team had decided on the exterior size.

The story of Toyota's Prius development highlights two aspects of innovation. First, this case showed how complex the product and the process of making it were. An automobile consists of 30,000 components and with the invention of a totally new model like the Prius, Toyota had to develop hundreds of new components. This in itself was a challenging process; moreover, the coordination of these components within the vehicle became an enormous task. A functional change in one component induced a series of changes to other components, and the development teams had to manage this snowballing effect. The best example in the case of the Prius was the battery development and its system reconfiguration.

Even after Matsushita Batteries successfully developed the batteries with 25 times more capacity, there were several critical changes in the overall design of the Prius, and the coordination among components was complex and iteratively evolving:

First, they realized that the originally proposed location for the batteries was inappropriate because they could not control the temperature of the batteries effectively. It was initially planned that the batteries would be located under the back seat, but the road surface temperature could easily change the battery temperature. Therefore, they changed battery location to the space between the back seat and the trunk.

Second, they found that this adjustment was still insufficient. Since it was critical to maintain the temperature of the batteries at the optimal, medium level for performance and durability, due to the memory effect, they had to add a dedicated cooling system for the batteries. These two changes involved major adjustments in the mechanics of the Prius, the cooling system, and the system integration of the battery operation. Figure 4.4 summarizes the changes made during development. We know what Toyota eventually decided for each of these elements. However,

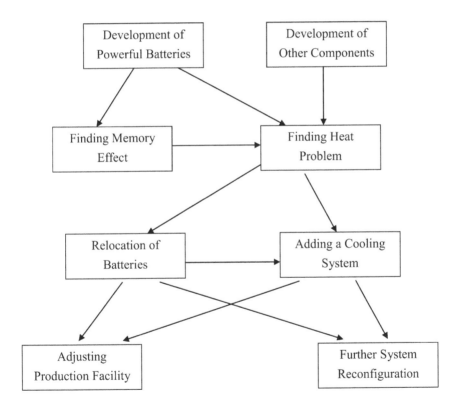

Figure 4.4   Induced changes in Prius battery development

Source: Author.

in each case, this was not a one-time decision made by the chief engineer or an executive. Instead, engineers in charge of these functions had to discuss options, such as where else to put the batteries, and analyze the pros and cons of each alternative over a lengthy time. This required the co-location of key players in all related divisions.

Furthermore, the changes in the development divisions were not the end. Since Toyota emphasized concurrent engineering to shorten the time lag between development and production, changes in components brought additional changes at the production facility. These changes included where, how, and in which order to assemble each component. As mentioned before, the configuration of the production assembly required input not only from the factory managers, but also from the development engineers. Co-location of all the staff involved was indispensable.

The other aspect of innovation making highlighted by the Prius story was the importance of prototyping and testing. Designing the system on paper or using computer software was only the first step, never the end of the development process. For instance, in order to scale down the engine space, which contains an engine, transaxle, damper, generator, planetary gears, decelerator, and motor, engineers from BR-VF and EV Development redesigned the size and shape of each component, tested its operation, analyzed the need for further rescaling, and tried again. Only after trying and testing could they devise a more effective space-saving configuration. In the end, they decided to abandon the conventional method of grouping these items, but took the decelerator out of the engine space and connected it adjacently.

The evaluation group played the critical role in testing and finding problems. The hybrid system was functioning theoretically with the batteries and engine, but testing revealed that the transition between the batteries and the gas engine operation was not smooth. In fact, it caused a physical shock like stalling. This result was fed back to the engine and battery divisions, as well as BR-VF, and they had to come up with a solution. It took almost a year to solve this problem, because they made two major modifications in the system: adding the adjustable air intake valve and controlling the air intake into the engine. Additionally, the evaluation group found a large noise problem with the relays that controlled different levels of electricity voltage. They solved this problem by covering the relays with sound-absorbing insulators and requesting that their component supplier add this extra process.

Each of the two aspects of innovation had significant implications in geographical terms. To manage the complex, snowballing aspect of the development project, their major development units had to be located in proximity, so that they could have frequent and intense coordination. This coordination was not as simple or hierarchical as having the chief engineer direct development units, but rather an iterative process in which all stakeholders provided ideas, discussed them, and tried to make improvements. Additionally, it was often multi-party coordination, and all of them needed to be available on any day, at any moment. To manage such complex coordination, co-location was the best solution. Moreover, the need for prototyping and testing required the co-location of the Prius and engineers. Whenever and whatever problems they found, engineers from potentially related divisions had to get together to analyze and identify the cause of the problems, propose possible solutions, test, and confirm the solutions.

Chapter 5

Canon's Bubble Jet Printer BJ-10v: The 13-Year Journey from Basic Research to Product

If you don't know the Canon culture, you cannot create a Canon product.
— Katsumi Masaki, Director, Global Technology Promotion Division,
Technology Management Headquarters, Canon (2004)

Canon developed a new printing method, called Bubble Jet, after much effort. It started with a small research group of five chemistry researchers in 1977 that grew to more than 200 in 1990. These project members moved from Meguro (1977–1979), Atsugi (1979–1983), Hiratsuka (1983–1989) to Tamagawa (1989–present). All these locations are in the southern Tokyo metropolitan area. More importantly, with each relocation all of the researchers moved together. This 'man transfer' was necessary because the development of the technology required the integration of various disciplines, notably physics, chemistry, electrical, mechanical, and simulation engineering. Co-location of the engineers and researchers was crucial for exchanging technical knowledge, suggesting new methods to overcome problems, and experimenting and improving persistently. All the major technological barriers were solved by combining these technical fields.

5.1 The Canon Corporation

The origins of Canon date back to 1933 when two camera lovers, Goro Yoshida and Saburo Uchida, set up a research lab in the hope of producing cameras of the same high quality as the then-dominant German Leica cameras. Although this was called a research lab, it was only a small room in a third-floor apartment at Roppongi, Tokyo. Hajime Mitarai, who was a medical doctor and later became the first president of Canon, financially supported the two researchers, who successfully created their original camera, Kwannon 1, in 1934. Yoshida chose this name for the first domestic 35 mm focal pressure camera because of his dedication to Buddhism. *Kwannon* means "Buddha" in Japanese, and it was later changed to *Canon* to be more accessible internationally (Canon 2005a). Canon Camera Inc., established in 1947, moved their headquarters and factory to the current location in Shimomaruko, Tokyo, in 1951.

Canon's principle of business diversification developed in the 1960s. Their strategy for entering the camera film business, still camera-related but requiring different materials and technologies, came about after a business trip to the U.S. by two executives: Hajime Mitarai and Keizo Yamaji, who became the company president in 1989. To explore sales opportunities in the U.S., they visited Eastman Kodak in Rochester, New York, in 1962. The two were surprised by their generous hospitality, and asked why. The answer was "Your cameras were film burners" (Nikkei Newspaper 1997). In other words, a camera manufacturer could make money only when a consumer bought their product. A film manufacturer, on the other hand, could continuously sell its products as long as consumers used their cameras. Furthermore, every time a film manufacturer changed its products, the camera manufacturer had to follow in order to be able to use the new film (Nakazawa, cited in Yoneyama 1996; 123ff). For this reason Canon decided to expand production into the nondurable film business.

In 1967, Mitarai announced a new slogan, "A camera for the right hand, a copier for the left hand" (Canon 2005a). The development of copiers was an application of recording technology to office equipment, and Canon intended to expand the business into both durable goods (the copiers) and non-durable goods such as cartridges (Sakakibara and Matsumoto 2004, 8). As the non-camera business division grew steadily, they renamed the company from Canon Camera Inc. to Canon Inc. in 1969.

Despite the economic recession in Japan since 1991, Canon has experienced continuous growth in each business both in and outside Japan, and sold 3.47 trillion yen (US$30.1 billion) and employed 108,257 staff worldwide on a consolidated basis in 2009. Sales of copy machines and printers surpassed its original camera business and accounted for 32.3 per cent and 33.2 per cent, respectively[1] (see Table 5.1). With its steady growth and international competitiveness, in June 2006 the then-president of Canon, Fujio Mitarai, became the chairman of *Keidanren*, a highly respected and politically powerful business association in Japan (Nikkei 2005). This appointment surprised the business world because this prestigious association historically selects its leaders from the traditional heavy industrial sector, such as steel, shipbuilding, and automobiles, not from newer industrial sectors, such as electronics.[2] This showed how perceptions of business leaders have recently changed and how well Canon has been regarded in Japan.

---

1   The resurging of the camera business, a jump from 11.8% in 2000 to 22.0% in 2004, was due to the growth of digital cameras.

2   Incidentally, the previous chairman of Keidanren was Hiroshi Okuda, a former chairman of Toyota.

Table 5.1   Consolidated sales of Canon products

	1993		2000		2004	
	bil. Yen	share	bil. Yen	share	bil. Yen	share
Cameras	180.4	10.2%	318.2	11.8%	763.1	22.0%
Office imaging products	643.4	36.3%	772.6	28.7%	1,121.0	32.3%
Computer peripherals	532.4	30.0%	1,023.0	37.9%	1,149.9	33.2%
Business information	306.6	17.3%	314.9	11.7%	117.1	3.4%
Optical and others	110.3	6.2%	267.8	9.9%	316.8	9.1%
Total	1,773.1	100.0%	2,696.5	100.0%	3,467.9	100.0%

Source: Canon, Annual Report, each year.

Canon is often compared to Sony because of its strong orientation toward the international market. Its original business intention was to produce Leica-level high-quality cameras and compete internationally. In its early years, soon after World War II, its main customers were U.S. military personnel stationed in Japan (Canon 2005b). Even after the occupation ended, Canon looked for a connection with the U.S. and established a branch in New York as early as 1955, five years earlier than Sony. It first sold cameras in the U.S. through Jardine and Matheson, a British trading firm, then through Bell and Howell, an American 8mm camera manufacturer. However, sales were not high, and Canon instigated direct sales in the U.S. in 1974. They launched AE-1, the world's first microcomputer-embedded, single-lens reflex camera, in 1976, and this product propelled Canon to the status of an international brand name (Nikkei 2001, 212–13). The manager of this early Canon U.S.A. operation was Fujio Mitarai.

Canon pursued internationalization not only through sales, but also by employing local production. In 1976, President Kaku announced the Premier Company Plan, which called for *kyosei*, a concept of "living and working together for the common good" at the global level (Canon 2005a). This specifically meant lowering tensions in bilateral trade by reducing exports from Japan and increasing local production in each market. Canon started overseas production sites as early as the 1970s: Taiwan (1970), Giessen, West Germany (1972), and Canon Business Machines in Los Angeles (1974). Other overseas factories were established in France (1983), Virginia (1985), and South Korea (1985). This was a very different pattern from other Japanese firms which started to expand overseas production after the Plaza Accord in 1985, resulting in rapid appreciation of the yen.

Moreover, Canon was an early developer of OEM among Japanese firms and outsourced calculator production to Calcomp, a Taiwanese manufacturer, in 1986 (Canon 2005b). As a result of its continuous international expansion, the overseas production rate increased from seven per cent in 1985 to 38 per cent in 2002 (Nikkan Kogyo Shimbun 2003, 110–11). The proportion of sales in Japan in 2004 was 24.5 per cent, which was considerably lower than their Japanese counterparts, such as Seiko Epson (46.9%) and Ricoh (53.6%). Although exact comparison is difficult, sales by U.S. competitors Xerox and Hewlett Packard (HP) tend to concentrate in North America, with some share in Europe, but less presence in Asia and the Pacific (see Table 5.2). In this sense, Canon is a global company with more efforts to reach every major market on the planet.

Table 5.2   Sales and geographical distribution of major printer companies

	Japan	Americas	Europe	Others	Sales in bil $
Canon	24.5%	30.5%	31.5%	13.4%	33.3
Epson	46.9%	16.4%	22.0%	14.6%	14.0
Ricoh	53.6%	17.9%	22.5%	5.9%	17.1
Xerox	-	53.1%	33.6%	13.3%	15.7

Source: Annual report of each company.

Note: Sales in billions of U.S. dollars.

Conversion rate: US$1 = 106 Yen.

5.2 Canon's R&D System

Canon is a highly research-intensive company in terms of both R&D expenditure and patent applications. In the fiscal year 2004, Canon spent 273,960 million yen (US$2.58 billion) in research and development, 7.9 per cent of total sales on a consolidated basis,[3] which included 184 companies worldwide (Canon 2004b, 33). Moreover, on an unconsolidated basis, this ratio was 12 per cent, while the industrial average of the precision machinery sector was 7.4 per cent. This figure was also significantly higher than the two previous case firms, Sony (7.5%) and Toyota (4.0%).[4]

Canon is an aggressive filer of patents at the international level. Each sector of Canon's core business, such as cameras, copiers, and printers, is research-intensive, and requires strategic development of technologies. Additionally, Canon files patents in platform and production technologies, such as fiber optics, semiconductor processes, software simulation, and evaluation methods. Canon has always been one of the major patent applicants in Japan and ranked second in

---

3   An unconsolidated basis means the parent company, in this case, Canon Inc. alone. A consolidated basis means figures for the business group, i.e. Canon Inc. and its subsidiaries combined.

4   This comparison on an unconsolidated basis may have limitations. Each company has a different organizational strategy and structure, which will affect affiliation of production and sales staff. For example, most of Canon's production and sales staff work at subsidiaries, while Toyota employs a large number of production personnel under Toyota Inc.

2004, after Matsushita Electric (JPO 2004). Canon has also been an active patent applicant in the U.S. (see Table 5.3), and it has been in the top three since the 1990s (USPTO 2006).

Table 5.3   Number of patents granted in the U.S. by organization

	1995		2000		2004	
	Company	No.	Company	No.	Company	No.
1	IBM	1,383	IBM	2,886	IBM	3,248
2	Canon	1,087	NEC	2,021	Matsushita	1,934
3	Motorola	1,012	Canon	1,890	Canon	1,805
4	NEC	1,005	Samsung	1,441	HP	1,775
5	Mitsubishi Elec.	973	Lucent	1,411	Micron	1,760
6	Toshiba	969	Sony	1,385	Samsung	1,604
7	Hitachi	910	Micron	1,304	Intel	1,601
8	Matsushita	854	Toshiba	1,232	Hitachi	1,514
9	Kodak	772	Motorola	1,196	Toshiba	1,310
10	General Elec.	758	Fujitsu	1,147	Sony	1,305

Source: U.S. Patent and Trademark Office (2006).

Around 6,800 researchers and engineers[5] conduct Canon's research and development activities, and 4,000 of them are located at Shimomaruko (Kozato 1999, 319), the firm's headquarters. Each researcher is expected to submit eight to twelve patent applications per year. The engineers and researchers are affiliated with the company at three organizational levels: (1) six R&D headquarters directly under the president, (2) development centers under each business unit, and (3) seven overseas R&D centers (see Table 5.4).

First, R&D headquarters hosts 1,800 researchers (Harryson 1998; 100). Normally, the scope of research within these six headquarters is on long-term, cross-cutting technological development, such as semiconductor processing and software development. Since the late 1990s, Canon has scaled down one of the branches, the Central Research Lab, which specializes in long-term, basic research and has been criticized for pursuing research unrelated to potential commercialization. Fujio Mitarai took the initiative on this restructuring, and downsized the number of researchers from 300 to 200, and this trend of scaling down on research operations is expected to proceed even further (Hara, interview May 4, 2004).

---

5   There are 18,000 employees under Canon Inc. (unconsolidated), which means 38 per cent of its employees are related to research and development.

Table 5.4   R&D-related organizations at Canon

Organization	Research Subject
R&D Headquarters	
Core Technology Development	Optics, simulation
Leading-Edge Technology Development	Semiconductor process, nano-tech
Platform Development	Software & security system
Device Development	Semiconductor devices
L Printer Development	Laser beam printers
Central Research Lab	Basic research
Business Units	
Image Communications	Cameras
Office Imaging	Copiers
Computer Peripherals	Printers
Ink-jet	
Device Development Center	
Chemical Component Center	
Chemical Engineering Center	
Chemical Recording Material Center	
Chemical Components	
Cartridge Development Center	
Optical	Fiber optics, semiconductor
Overseas R&D Centers (7)	

Source: Canon (2006c).

Note: Italics are Bubble Jet printer-related research units in 2005; therefore, organizational names do not precisely match with division names for the development of the BJ-10v, the case product in this chapter.

Second, development centers under business units employ the majority, approximately 4,200 or 62 per cent, of the research workforce. Their primary target is commercialization of products, usually within a three-year term (Hajimoto, interview, July 20, 2004). Additionally, 80–90 per cent of Canon's patents are related to production technology, and this is where most patent filings take place (Toyosaki, cited in Harryson 1998, 102, 24).

Third, Canon operates seven overseas R&D centers and employs over 800 staff (see Table 5.5). With the 'One-Lab, One-Concept' policy, each center is expected to be uniquely positioned so that it does not overlap with other Canon R&D centers. As a result, the overseas R&D centers focus on software development and applications. In theory, these areas complement the hardware expertise of

R&D headquarters in Japan (Kozato 1999, 324). In order to share and exchange information, the staff of the overseas R&D centers and a small portion of Japanese researchers organize global meetings twice a year, in addition to ad-hoc, project-based needs (Hajimoto, interview, July 20, 2004).

Table 5.5   Canon's overseas R&D centers

Name	Location	Year	Emp.	Function
Development Americas	Irvine, CA	1990	117	Networking, printing, imaging, optics
U.S. Life Sciences	Arlington, VA	2002	5	Diagnostics, medical instrumentation
Technology Europe	Berkshire, U.K.	1988	49	Information interaction, mobile
Research Centre	France	1990	59	Home-network, imaging & web
Information Systems	Sydney	1990	251	Digital imaging, applications
Information Technology	Beijing	1998	129	Language processing, e-learning
Information Technologies	Philippines	1991	221	Application equipment, software

Source: Canon (2005a).

5.3 Overview of Printer Technology

This section will describe different types of printer technologies. The objective is, first, to explain how complex and technology-intensive they are and second, to show how complexity forces each firm to establish entry barriers, most commonly using patents, and to institute a proprietary style. Understanding these factors and major technological fields can explain Canon's rationale when searching for a new printing method—Bubble Jet technology.

Earlier printers used an impact method, in which a printer physically hit paper with a hammer. This consequently caused noise problems especially in the office environment and had significant limitations regarding resolution levels. As an alternative, non-impact methods were developed—the first was called thermal printing. These printers melted plastic ribbons using heat and transferred the ink onto the paper. They considerably reduced noise levels and improved resolution levels. However, thermal printing had high running costs: For any letter printed, "." (period), a given space on the ink ribbon had to be abandoned, and a new space used for the next letter. Moreover, it required special thermal recording paper, and

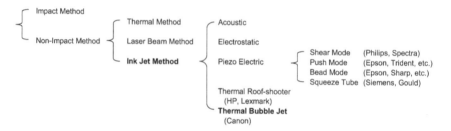

Figure 5.1   **Types of printing technologies**

Source: Based on Le (1998) and Yoneyama (1996), and modified by the author.

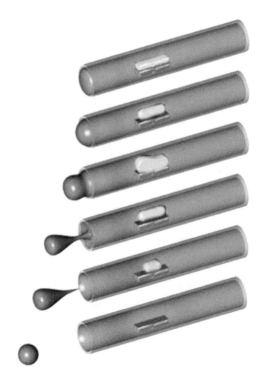

Figure 5.2   **Process of Bubble Jet printing**

Source: Canon (2006e).

the melting process was slow. Therefore, it did not emerge as a major printing method, and these two printing methods became virtually extinct in the 1990s.

What emerged in the late 1980s and still remains as one of the major technologies today was the laser beam method. This method was applied from technologies involved in copier machines and used the following processes. First, an electrical charging roller negatively charges a photosensitive drum. Second, laser beams are emitted and reflected through a polygonal mirror onto the photosensitive drum. Third, toner is affixed only to the areas negatively discharged by the beams. Fourth, paper is attached to the drum, and this time a positive charge is applied to attract toner to the paper. Finally, heat and pressure is applied to fix the image onto the paper (Canon 2006b).

An increasingly popular alternative method was the ink jet, which is a piezoelectric method. This system used piezo-crystals that changed their shape depending on the electric charge applied. The difference between a positive or negative charge would inject an ink droplet onto the paper. Philips, Epson, Sharp, and Siemens were major companies applying this method.

Within the ink jet method, Canon searched for a further alternative and developed the Bubble Jet. With this, first, a heater attached to a printer nozzle was electrified and produced heat. Second, this heat vaporized ink and produced a bubble inside the nozzle, which pushed out an ink droplet. Third, the bubble shrank and prepared for the next movement (see Figure 5.2 for each step in the process).

The Bubble Jet printer was less complex mechanically than the laser beam printer and could be produced at lower cost and on a smaller scale. Its advantages included color printing and the use of normal paper. At the same time, it was a highly technologically intensive method, involving a wide range of fields such as physics, chemistry, precision machinery, and electrical engineering. More specifically, it required chemistry for ink liquid formation, physics and hydrodynamics for ink flow, ultrasound and the electrically-charged particles for ink droplet formation. (Sakakibara 1998, 9–10). Canon introduced the popular BJ-10v model in 1990, and consequently pushed its market share in ink jet printers over 80 per cent in Japan in 1991. Although this overwhelming dominance was eroded in the mid 1990s when other firms introduced alternative models, Canon is still one of the top three printer manufacturers in the world. Canon has filed more than 3,000 patents related to this printing method up to today (Ishikawa, cited in Yoneyama 1996, 132).

5.4 Development of the Bubble Jet Printer: BJ-10v

Development of the Bubble Jet printer was a long, persistent trial and error process through an inter-disciplinary technological approach. In the early stages, during the initial search for suitable technology, it was an attempt to integrate mainly chemistry and physics processes. As Canon realized the potential of Bubble Jet technology, during the component and concept development phases, they stepped up for commercialization by incorporating elements of electric and mechanical engineering. Later, they applied semiconductor processing and simulation engineering to improve the Bubble Jet printer.

5.4.1 The Search for Technology

The early years of Bubble Jet Printer development were a difficult period as the team at Canon looked for an alternative printing technology from scratch. Even though the basic concept of using a bubble to push ink was a breakthrough idea, it met tremendous technological obstacles. The project started with a small team of five people at a university laboratory in Meguro and gradually increased to 26 members at Atsugi, Canon's Central Research Lab. Overcoming each problem was a lengthy process, and the integration of various fields was essential. Canon gathered experts in materials science, as well as electrical and machinery engineering.

Canon launched its first laser beam printers in 1965 by applying its own New Process method, which was developed as an alternative to the dominant Xerox method. However, this business did not grow as expected, and the laser beam printer (LBP) technology had three limitations. First, the structure of LBP was so complex that printers had to be very large. With so many complicated components, they could not be produced any smaller nor more cheaply. Second, despite its high quality black and white printing, LBP was not appropriate for color printing (Yoneyama 1996, 130–31). Third, the growth of the market already had started to slow down even in the 1970s. As more competitors came into the market, Canon expected LBP to be less profitable in the long term.

It was over these concerns that Keizo Yamaji, a director of the Copier Development Division, agreed with Hiroshi Tanaka, a senior researcher, and ordered a prominent researcher, Ichiro Endo, to look for an alternative printing technology in the mid 1970s (Endo 1993, 17–18). Incidentally, this Yamaji, Tanaka, and Endo team continued as a strong backbone of Bubble Jet printer development and later, all three became major figures at Canon: Yamaji as a president of Canon in 1989, Tanaka as a vice-chairman, and Endo as a senior executive board member (Canon 2005a). Endo, trained as a chemist, led a team of five physics and chemistry researchers and started to conduct experiments in a laboratory at the Tokyo Institute of Technology, a public university (Omata, interview, September 24, 2004). It was located at Meguro, half an hour from Canon's headquarters at Shimomaruko (see Figure 5.3).

Endo started by researching previously patented recording technologies. There had been more than a few patents already filed, and surprisingly the principle of an ink-jet method had been proposed in the U.K. as early as in 1879. Between the 1950s and early 1970s, American and Swedish scientists filed eight patents. Based on these inventions, IBM, Hitachi, Matsushita, and Xerox started to move towards commercialization (Iwai 1997, 125–6). Endo explains: "In early 1977, I was looking for an alternative energy source other than the piezoelectric device and tried every possible one, such as light and electricity. Then, I realized there was no patent with thermal energy and thought, what about combining the thermal and the inkjet? In July, I experimented with an injector, and it came out as rapidly as 10 m per second" (Endo, cited in CMRG 1995, 87).

The first intra-firm linkage in this product development happened as Endo procured potential testing devices from other divisions within Canon. It was a random connection. At that time Canon's calculator division was experimenting with a combination of thermal printing and a calculator. Endo imagined that the printer's thermal technology could be used as an energy source and requested a thermal head from them. Additionally, he came up with the idea of using optical

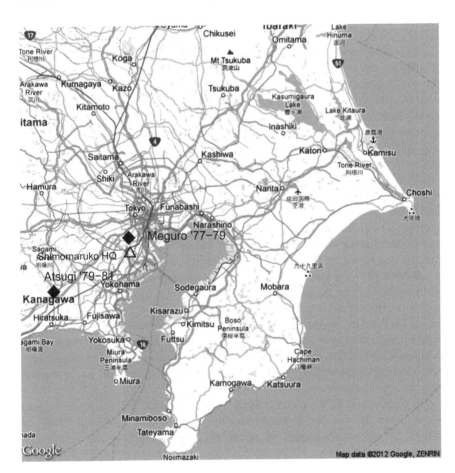

Figure 5.3   Map of BJ development 1: 1977–81

Source: Author.

fiber and contacted a fiber research group. Normally, an optical fiber was a string without holes, but Endo made a special request for a fiber with a hollow center of 100 micrometers. He inserted ink into the optical fiber through capillary action and pressed the ink through the tube with a thermal head. This method was a breakthrough, and Endo filed a patent on October 3, 1977 (Miyazaki 2002, 166–7).

In response to this discovery of potential new technology, Hajime Mitarai, then a director of the Product Technology Center and later the president of Canon (1993–1995), authorized a task force called Canon Inkjet Components (CIC) in April 1978 (Yoneyama 1996, 134). The five researchers moved to the Central Research Lab at Atsugi, 45 km or 28 miles southwest of the company headquarters at Shimomaruko. In addition to analyzing inkjet printers manufactued by other companies, the team researched the feasibility of development in three key areas: ink, the system, and thermal jet technology (Taniishi, interview, September 24, 2004).

In general, new technology or product development is a probability business. In other words, a few new ideas succeed, while many of them do not. Especially at the early stage of new-technology development in which the probability of success was low, the establishment of a formal division with staff could be costly and could create inflexibility in the course of the technology and product development, because once a formal organization is established, it tends to resist any change in a development project. The "task force" was an organizational practice typical at Canon which introduced it to this practice with the intention of making it less costly and inflexible. Through this method, the research team members were able to maintain formal organizational affiliation to their original divisions, yet still participated in this project. The members belonged to an inkjet development division, a calculator development division, and a production technology division, and all these units were under the Central Research Lab. They were assigned a specific task to last for three years (Omata, interview, September 24, 2004).

In April 1979, the task force focused its objective on developing a thermal head for the printer. Here, the nozzle of the printer head was causing a major technical problem—the ink did not come out as designed because of heated bubbles in the head. The bubbles lowered the pressure in the ink and caused congestion in the liquid flow at the top of the head. To come up with a method to pull out bubbles efficiently and effectively, they added experts in chemistry, physics, mechanical and electrical engineering (Endo 1993, 20). They did not find an immediate solution, and the problem persisted for several years.

In September 1979, a new task force, Component and Inkjet (C-I), was formed, and the number of researchers increased to 26 (Yoneyama 1996). At around this stage of the project, the members suffered from three additional major technological difficulties: (1) cogation, (2) passivation, and (3) unequal injection of ink. Each required a different solution and mobilized a different combination of technical experts.

First and most problematically, cogation was a problem such that the ink got burnt and caused further congestion inside the nozzle. The word *coge* means burnt deposit or an act of getting burnt in Japanese. Endo searched for a technical term for this phenomenon, but did not find anything in English, so he decided to create a new term, *cogation*, which is internationally used today. As explained, the thermal system heated ink to create a bubble. Ink dyes could normally resist heat up to 250°C; however, the temperature in the Bubble Jet system would momentarily go up to 300°C. Endo and his team visited the Tokyo Fire Department to hear a lecture on the process of cogation and heard from professors specializing in pigments. Then, they tried every possible type of dye. Experts in materials science with a background in crystal display, microelectronics, and machinery worked on this experiment (Endo 1994, 34). This was an extremely lengthy process. They heated polyethylene glycol solution to 200°C, dissolved the dye, and put it through the heating process. By the end, they had tried about one thousand types of dye. They finally found some non-burning dyes and applied them to create new types of ink.

The second problem they faced was passivation. Water-based ink could easily cause electric insulation, which damaged one of the heater electrodes. Chemical experts researched various coating processes and methods, and eventually solved the problem by inserting an insulating film between the heater and the ink (Endo 1994, 20).

The last problem was unstable injection of the ink. This was a complicated problem exacerbated by the previous problems. First, cogation caused congestion inside the nozzle. Additionally, the heater was heating unevenly due to passivation. Therefore, the power level of the heater was not consistent, and each bubble created differed in size. As a result, the ink came out unevenly. This problem persisted for a long time, even through the 1980s, and the Bubble Jet development team made a series of improvements. They tackled the problem by integrating materials science, semiconductor processing technology, and precision technology. The semiconductor processing and precision technology came from the technological expertise of other business divisions at Canon. The ideas came from researchers in the Bubble Jet taskforce who had previously worked in the semiconductor and optics fields.

Overcoming these major technological difficulties, the task force came up with a prototype which was presented at the Canon Grand Fair in November 1981. This was an exhibition in which each Canon's business division would typically display their existing products, but not usually research-in-progress items. The senior executive, Yamaji, requested that Endo use this exhibition for a political reason. Up until this stage, the Bubble Jet had not received much attention inside Canon. In fact, many researchers were skeptical whether this technology would be commercially viable in the near or even in the far future. In this sense, two senior figures, Mitarai and Yamaji himself, were exceptions. They thought it would be necessary to present a prototype and demonstrate the feasibility of the technology and product in order to acquire organizational support at the company level. Though Endo was aware of limitations in the Bubble Jet technology at this point, he swiftly agreed. Moreover, Endo thought that setting a concrete goal in the short term, rather than aiming for commercialization in ten years, was essential to motivate the project members (Endo, in Iwai 1997, 144).

The presentation at the exhibition turned out to be more than successful. The section exhibiting the Bubble Jet printer was always full of people and competitors in the same industry, such as Epson and Hewlett Packard (HP), as well as potential customers showed a great deal of interest in the technology. This exhibit had two consequences. First and most importantly, the Bubble Jet printer got its first customer. A president of Schlumberger, a French oil field drilling company, was impressed with the printer and requested development of high-speed color graphic recorders for petroleum search activities (Endo 1994, 34; Miyazaki 2002, 168). Although the volume of this order was relatively small, 2,000, and the development team still had to make substantial changes for the specific use, this news convinced the team members that their product had commercial potential. Additionally, other people at Canon had to accept that this would be a viable product for the company in the future. The official order came in the summer of 1983, and Yamaji, Mitarai, and Endo won political support for the project.

In addition, HP approached Canon about a possible collaborative technological alliance. Not surprisingly, HP was developing similar technology. Canon happened to file a base patent only a month earlier than HP and came up with a publicly presentable model sooner. This discussion of an alliance first started at a personal level: Hajime Mitarai, then the director of Canon Central Research Lab, had received his doctoral degree in engineering from Stanford University, and was a good friend of a director of HP's Research Lab, also located in Palo Alto, California. In October 1982, two of the leading companies in the printer industry started to exchange information (Endo 1993, 21). Eventually, they terminated this official alliance by 1985 as a result of strategic differences and pursued technologies and products of their own. HP introduced ThinkJet models in 1984 (Le 1998). Nonetheless, this built a platform for cooperation between the two companies on a broader scale. For example, it resulted in a technological alliance in the semiconductor sector in the mid 1980s (Canon 2005b).

5.4.2 Component Development

The period of component development, roughly from 1983 to 1987, saw a gradual scaling up of the development team devoted to commercialization of the printer as a product. Further electrical and mechanical engineers joined the project and when the teams were relocated from Atsugi to Hiratsuka, the location of Canon's Component Development Center, it should be noted that all the staff moved together. One major technical problem was solved serendipitously through Canon's camera technology, precision machinery, and photo recording technique. Still, an integration of different technical fields, especially in engineering and physics, played a critical role in development.

The 1981 exhibition demonstrated that Bubble Jet technology was technologically and commercial feasible, if not ready yet. Even the world's leader in the printer industry, HP, evaluated this technology highly and asked for collaborative research. Now, the whole organization of Canon acknowledged that the task force had established the major fundamentals of Bubble Jet technology; the next step was to apply it for commercialization. Consequently, they reorganized the development workforce by up-scaling the project.

In August 1983, they launched the next phase, I-Task. "I" stands for Inkjet, as one might expect. Additionally, the pronunciation of "I" (as ai) is the same as love in Japanese. A project member joked that "The project was operated by people who officially and proudly loved Bubble Jet" (Ohta, cited in Iwai 1997, 160). The objective now shifted from developing technology at whatever cost and using whatever resources necessary to developing a product at a commercially reasonable cost (Endo 1993, 22).

Spatially and organizationally, there was an important change. First, they mobilized production technology staff and increased the total project members from around 50 in 1983 to 120 by 1985 (Yoneyama 1996, 137). The task force split into two divisions: one for functional material development, headed by Endo, and another for functional device development, led by Watanabe.

Geographically, these development teams moved from Atsugi to Hiratsuka, 52 km or 32.5 miles southwest of the Shimomaruko headquarters and 21 km or 13 miles southeast of Atsugi. The I-Task had formerly been under two divisions: the Central Research Lab and the Technical Component Center. It was Canon's own way of bridging the research and development process by gradually shifting its intra-organizational affiliations. More importantly, Canon moved all the researchers and engineers involved from Atsugi to Hiratsuka. Conventionally, when large high-tech firms transfer the base technology to commercialization, they transfer technology from research labs to business units, but not usually the staff of the research labs. That is, it was normally a technology transfer system in a product development process. In contrast, Canon distinctly pursued a "man transfer" system and moved staff and technology together (Taniishi, interview, September 24, 2004).

In mid 1985, Canon was getting close to introducing the BJ-80, its first mass-market Bubble Jet printer, into the market. However, a major problem emerged. They found that the heaters in the printers had low durability and broke for unknown reasons. To investigate the root of the problem, they mobilized not only all the staff under I-Task, but also all junior engineers at the Component Development Center. It took a while, but they found the problem and named it *cabitation*. Once again, it was the employment of a variety of business and technological fields by Canon that enabled analysis of this problem. Photo-recording technology, a core technology under Canon's original camera business, was the key. Ikeda, an engineer, was investigating this problem using a microscope with a high-speed recording function and he observed that bubbles were coming out of the heater, while they were supposed to come out only from the nozzle. This had happened because a burst of bubbles in the nozzle had created a vacuum, and this shock eventually damaged the heater (Endo 1993, 20). The solution they came up with was to insert a protecting film between the heater and the nozzle. This was a matter of physics, and they tried various materials that would withstand the vacuum shock. It took four months to fix the problem and restart production and sales (Taniishi, interview, September 24, 2004).

Eventually, an experimental model, BJ-80 (in Figure 5.4), came out in November 1985. This printer received mixed evaluations—despite some strengths Canon had to acknowledge the need for further improvements. While the printer was equipped with new technology, its exterior body was the same as that of Canon's existing laser beam printers so there was no major impact through its physical appearance. Nonetheless, the development team was proud of its high quality when compared with other products on the market. In terms of printing quality, the BJ-80 had 24 nozzles, twice as many as HP's existing printers, which yielded double the resolution quality, 180 dots per inch. The durability of the printer head was approximately ten times longer than conventional dot matrix printers and several hundred times better than HP's ink jet printers[6] (Omata, interview, September 24, 2004). Optimistically, several project members conducted a marketing tour in the U.S. where they visited DEC, IBM, and three engineering academic conferences, and the tour turned out to be successful. Although the U.S. market, at that time, was paying more attention to the piezo-crystal method, IBM showed an interest in Bubble Jet technology and later requested OEM production from Canon.

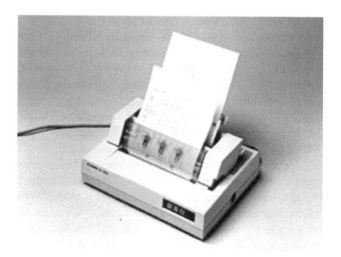

Figure 5.4   BJ-80 in 1985: The first Bubble Jet printer, but still with a traditional printer style

Source: Canon (2005b).

---

6   Ink jet printers of this period generally had only short durability.

On the other hand, Canon U.S.A., a sales arm of Canon in the U.S., provided critical feedback from a marketing perspective when they suggested developing printing methods that would work with normal paper, instead of specially coated paper. So far, I-Task had not yet developed inks that were compatible with normal paper. I-Task members promoted three advantages of the BJ-80: that it was quiet, fast, and printed at high quality. While Canon U.S.A. acknowledged these advantages, they critically demonstrated that the BJ-80 was superior as a manufactured good, but not sufficiently good to be a marketable product yet. The I-Task members returned to Japan with the urgent goal of making improvements toward a marketable *product*.

5.4.3 Concept Development

During the concept development phase, Canon attempted to improve the Bubble Jet technology as a commercial product. In order to do this, three development processes required co-location of the participating engineers and researchers with the goal of integrating the technology with a commercially appealing product. This involved (1) brainstorming for new concept development, (2) the integration of engineering and chemistry disciplines for further technological advancement, and (3) the creation of prototypes to test both the technology and the final product. This section will discuss each of the three aspects in detail with examples.

Canon's Bubble Jet printer development took a different path from the two other product cases in this book. In the development of the Vaio 505 and the Prius, both Sony and Toyota first started with a product concept, then searched for available and matching technology for the product. In contrast, Canon developed the technology first, then searched for a concept to reconfigure the product. These differences show that new development in modern consumer products involves different types of innovation, but there is no universal order to which types of innovation must come first. In other words, it could be a concept generating a product and calling for the necessary technology, or technology generating a product and calling for concepts based on marketing perspectives. Innovations may come from multiple directions; innovations can diffuse in multiple directions.

Up until this stage, development of the Bubble Jet printer involved engineers in the development of core and production technology. In general, these staff members were engineers, who tended to perceive technology as a driver of products (Omata interview, September 24, 2004). After realizing the potential market size and the need to make improvements from marketing-based perspectives, such as using normal paper types suggested by the U.S. subsidiary, Canon executives decided that Bubble Jet products needed product configuration.

Yamaji, a senior executive, assigned Tanaka to take on the responsibility. This appointment was a major organizational upgrade because Tanaka was the executive in charge of Office Equipment Business, the largest and most profitable business sector of Canon at that time. Tanaka identified the next stage in commercialization of the Bubble Jet as a high priority for the firm and mobilized the directors of major recording technology divisions (Ohmori, in Yoneyama 1996, 138–9). Hence, the B-Project, standing for "Business," started in July 1987. The project members moved from the Component Center at Hiratsuka to be under the Office Equipment Business at Tamagawa, a town next to the headquarters in Shimomaruko. The number of members grew to 200 full-time and about 100 part-time engineers (Omata, interview, September 24, 2004).

A major figure in generating marketing concepts was Saito, a director of the Peripheral Component Development Center,[7] which was the driver of laser beam printer development, a rising business unit at Canon. Saito had two years' experiences at Canon U.S.A. where he was exposed to the sophisticated marketing concepts developed by HP and other U.S. firms (Iwai 1997, 189–90). He quickly saw that the Bubble Jet was outstanding technologically, but it had been developed mainly through functional engineering improvements. The strengths of the Bubble Jet were not fully expressed in the products, and there was little visible difference from other types of printer, such as dot matrix and laser printers. The originality of Canon's own technology needed to be identified and expressed.

Saito initiated a series of brainstorming exercises for concept making. Unlike Sony's brainstorming exercise, which involved ten selected members, Saito arranged it on a much larger organizational scale. He required every division of the Peripheral Component Development Center, more than 30 staff, to form its own brainstorming group. Then he aggregated and analyzed all the inputs. Some divisions had a boot camp method in which they spent three straight days in one room for this exercise (Saito, in Miyazaki 2002, 171–2). This concept making was a process of presenting a potential product concept, criticizing, modifying, and often scratching and starting it over again. Such information must be transacted instantly and required face-to-face communication, in which all members had to be present in one room.

Before finalizing a product concept, they initially analyzed two aspects: the strengths of the Bubble Jet and its current market use. First, in terms of the strengths, the Bubble Jet had three robust advantages over dot matrix printers: quietness, speed, and high quality. However, the Bubble Jet was not superior to laser printers, at least in speed and quality. On the other hand, laser printers had more complex mechanics, which resulted in a large and costly structure. Additionally, they were not suited to color printing. Therefore, the Bubble Jet had potential for smaller, cheaper, and color printing (Sakakibara and Matsumoto 2004, 11). Regarding the marketing analysis at the time, printers for PCs were mainly for office use, with high volume and high speed. The cost of laser printers was as low as $1,500, while that of dot matrix was about $1,000. Nonetheless, they considered that this price was still relatively high. They reached the conclusion that, with its compact structure, the Bubble Jet printer had the potential to be marketed in a much cheaper price range.

---

7   This Peripheral Component Development Center was located at Tamagawa.

Along with this compactness and low-price advantage, the development team considered mechanical integration of a printer head and ink tank. At the time, the durability of the printer head was an issue, and producing a permanent head was still costly and technologically demanding. Durability would depend on customer use, but the head would still require regular maintenance. In response, Canon, like its copier and printer rivals, established business service centers and provided maintenance services on the printer head for business customers. This opened up new business opportunities, and became a dominant and profitable business model for the printer and copier industry of today. However, for targeting cheaper price ranges, i.e. for households, providing maintenance services would be inefficient and unprofitable. They had to come up with a printer head so that would enable any household to use a printer without needing major maintenance.

A solution came from Kondo, a senior executive, who proposed the integration of the printer head and ink tank; that is, to use the printer head as a disposable component and replace it when the ink ran out. In this way, all they had to provide was a printer head with limited durability. Moreover, this disposable printer head could be substantially cheaper than a permanent printer head. While this would be an additional technical challenge in development, that would also mean a higher technological barrier for competitors wishing to imitate the technology (Sakakibara and Matsumoto 2004, 6). This became the business model for the printing industry today.

Figure 5.5   BJ-10v: compact and portable

Source: Canon (2005b).

Adding a product concept to the available technology required further technological development. This section will introduce two examples of how Canon overcame such technological barriers by the amalgamation of several technological fields: (1) integration of the printer head and ink tank, and (2) acceleration of printing by using mechanical and simulation engineering.

First, integration of the printer head and ink tank came through semiconductor processing, a field requiring knowledge of both chemistry and engineering. They had been using glass for the printer head, but looked for alternative materials in order to produce it more cheaply. While experimenting with materials, they also developed a production method that created a narrow hole in the printer head. Making a hole in an odd shape, such as a trapezoid, was relatively easy, but making a round hole was problematical, and they tested several methods, such as using a press perforator, a carbonated gas laser, etc. Each machine cost the equivalent of several million U.S. dollars, but the senior executive approved them instantly and in the end, they developed a printer head using integral molding plastics and succeeded in making the holes with an excimer laser (Yoneyama 1996, 139; Iwai 1997, 188–9).

Simply put, they decided not to use a thin pipe as a nozzle, but developed special plastics and generated extra-thin holes to function as nozzles. First, they placed heaters on a silicon wafer, and covered them with a layer of special resin "A". Using a photolithography machine, the layer covered with resin was exposed and solidified. Next, they created an additional layer using another kind of special resin, "B". Similarly, a photolithography machine was used to solidify the B layer. Finally, using precision technology, they created injection holes on the top, ink-supplying holes on the bottom, and removed the "A" layer. Innovation in this production system omitted all conventional bonding processes and allowed for more precise mechanics, which was essential to achieving higher resolution. Moreover, Canon developed this process by integrating three areas of technology: (1) chemistry—creating two kinds of special resins, (2) precision technology—creating thin holes in the right place, and (3) semiconductor processing—using silicon wafers and photolithography machines.

The second example of the integration of technological fields was the acceleration of printing speed by applying simulation engineering to electrical and mechanical engineering. As described in Figure 5.2 (above), in a Bubble Jet printer, ink would come out of the printer nozzle as a result of being pushed by a bubble. Because the shape of a liquid ink droplet changes over time, the surface of the ink under the nozzle would fluctuate. Here, timing was an issue that significantly affected the ink diffusion of the next round. If the next bubble came out too soon, the volume of the ink droplet in the next round was different, which resulted in uneven printing. Although an easy solution could be to wait until the surface of the ink stabilized, which would eventually happen after the total separation of an ink droplet, the waiting period would slow the printing speed considerably. Thus, they needed to calculate (1) how and when an ink droplet was discharged, (2) how the surface fluctuated, and (3) when the next optimal timing would be.

**Figure 5.6   The changing shape of an ink droplet, estimated by simulation
method**

Source: Canon (2006a).

This required theoretical knowledge in hydrodynamics and the use of simulation technology, which was originally developed by HP but acquired by Canon through the technological alliance in the mid 1980s. Asai, who studied physics in college, was assigned to this task, and eventually developed a method of pushing out the ink 300 times per second (Taniishi, interview, September 24, 2004).

Lastly, prototyping was essential to complete the development process, and it required the physical presence of engineers to analyze the functions and attractiveness of the product. This was where information technology facilitated the development process, yet it could not be applied for specific aspects of development of the product interface. After development of the base technology, "engineers had to apply as many of the five human senses as possible for product development," recalled Taniishi, then a project member and currently a staff manager at the Core Technology Development Headquarters (interview, September 24, 2004). The engineers always had to think about how consumers would feel when a product was sitting or functioning next to them. For a Bubble Jet printer, these considerations included more than a few dimensions: the physical appearance (size, color, shape), weight, speed and quality of printing, noise levels, etc. Information Technology could not simulate consumers' feelings about many of these features. Although Canon had always been a leading adapter of CAD, which could present a digital image on a computer, it could not predict the potential reactions of consumers. What a consumer might feel when looking at two- or three-dimension digital images on a computer screen was irrelevant—it must be an actual, physical printer to look at. Additionally, a person had to touch and carry it in order to have a tactile response. A description of 1.8 kg (4 lbs) was merely empty advertising unless one actually lifted a printer and felt how light it was. Furthermore, there was no noise simulation technology—someone had to start and listen to an actual printer performing. All this meant that the engineers had to develop prototypes and test every single feature. Only after this process, could they gather feedback and make modifications to the product.

5.4.4 Toward Mass Production

Initially, the production facility for Bubble Jet printers was located at Tamagawa with the development division. As the BJ-10v became accepted in the market and production volume rose, Canon mobilized two factories outside the Tokyo metropolitan region: Ami in Ibaragi prefecture and Fukushima.

Toward the end of the 1980s, Canon was conducting a major organizational restructuring, which affected the production of Bubble Jet printers. In 1988, Canon celebrated the 50th anniversary since its establishment in 1937. During this period, the growth and core business was based on camera products, but the company further committed to multi-business operations. Thus, President Kaku declared the year 1988 as the Second Establishment of the Company and called for a global company initiative.[8] As part of this restructuring, Canon removed camera production facilities from Tamagawa, replacing it with the development divisions of cameras and printers, including the Bubble Jet. Now, Canon concentrated all the development and production units of the Bubble Jet in one place.

As the scale of Bubble Jet production increased, they relocated production to Ami, Ibaragi. This was on the eastern side of the Tokyo metropolitan area, 87 km (54 miles) from the headquarters in Shimomaruko. While there were 600 production personnel at the Ami factory, development and production technology staff from Tamagawa visited Ami frequently, sometimes every day, between 1989 and 1990 (Taniishi, interview, September 24, 2004). The official production of the BJ-10v started in July, and it went on sale in October 1990. This model was drastically cheaper than other ink jet printers, 74,800 yen (US$554).[9] HP tried to lower its price and introduced a new, cheaper model in 1991; however, it was still $1,100.

The impact of the BJ-10v on the printer market was phenomenal. At the onset of BJ-10v sales, Canon was one of the leading players in Japan, but tied with Seiko-Epson at around 30 per cent of the ink jet market. During 1991, Canon's market share jumped to 80 per cent, and this dominant position continued until 1995, when Seiko-Epson introduced alternative models (Yoneyama 1996, 141).

As the sales of Bubble Jet printers surged, Canon added the Fukushima Factory to the production line. This factory was located three hours north of the Ami Factory and formerly produced camera products. More than 1,300 employees worked at this factory (Canon 2004b).

---

8   The principle idea of this initiative was mutual development based on a collaboration between different countries. At the operational level, Canon established nine overseas research labs in the following five years and promoted local procurement from 30 per cent to 54 per cent (Nikkei 2001). For example, in this year alone, they established Canon Research Center Europe (UK) and agreed to a strategic alliance with Beijing University (http://web.canon.jp/about/history/main09.html).

9   The exchange rate at the time was US$1 = 135.40 yen (Bank of Japan). http://www.boj.or.jp/theme/research/stat/market/forex/index.htm.

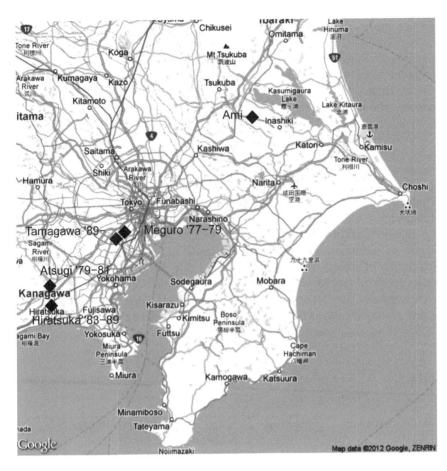

Figure 5.7   Map of Bubble Jet development 2: 1977–2004

Source: Author.

The geography of production went to a cross-border scale. Canon established a subsidiary in Thailand in 1990 and assigned Tanigawa, a veteran production manager, to this mission, along with 25 other specialists in production technology, quality assurance, molding, materials, etc (Iwai 1997, 48, 51). Simultaneously, the Thai factory sent key production workers to the Tama, Ami, and Fukushima factories for training. The production operation started with 360 workers in 1992.

5.5 Summary

Canon's development of Bubble Jet products was a long and arduous journey. From the filing of the first patent was in 1977, it took 13 years until the popular BJ-10v came out at the end of 1990. It started with about ten years of basic research and technological development. During this period, development took place in the southern Tokyo metropolitan area, and the team moved from Meguro (5 employees), to Atsugi (50), and then to Hiratsuka (120). These development centers were located within one and a half hour's distance from headquarters at Shimomaruko. By 1987, Canon was committed to commercializing this technology and mobilized more staff (200) at Tamagawa, a town next to the headquarters. Production first started at Ami, Ibaragi, in 1990, then spread to Fukushima and Thailand.

The story of the BJ-10v emphasizes the interdisciplinary and serendipitous nature of technological advancement because of its intense and systematic integration of various technological fields. Table 5.6 summarizes the kind of major technical problems Canon faced, how they were solved, and what kind of technical fields were involved in the problem solving. Since Endo had found the use of optical fiber and a thermal instrument for the printing mechanics, his team faced at least seven major technical limitations.

Table 5.6   Seven problems faced and the technological fields used to solve them

Major Limitations	Solutions	Integrated Fields
Cogation	Finding a heat-resistant dye, adding polyethylene glycol	Materials science, chemistry, microelectronics
Passivation	Coating to insulate a film	Chemistry
Heater durability	Using photomicrographs, adding a thin film	Optics, machinery, chemistry
Limited paper use	Finding a new type of paper	Marketing
New market opportunity	Cheap and color printing	Marketing
Durability of a printer head	Integrating the head and ink tank, using plastics and lasers	Semiconductor processing, precision machinery
Slow printing speed	Applying hydrodynamics, calculating ink liquid fluctuation	Chemistry, simulation engineering

Source: Author.

Cogation, the burning of deposits of dried ink as a result of the high temperature needed to create bubbles, was the first major problem. They tested around a thousand types of dye to identify the most heat-resistant one. Additionally, they discovered a method of dissolving dye into the heated polyethylene glycol solution, which also created higher heat resistance. This was a combination of materials science, chemistry, and microelectronics.

Similarly, they solved six other technical problems by trying new methods from different technical fields, starting with physics, chemistry, and electrical engineering. They soon added optics, semiconductor processing, precision machinery, and simulation engineering. In the later phase of the project, marketing perspectives were also employed as viable inputs for expanding the market scope of the BJ-10v through paper type, color, and cheaper printing.

Each technical problem was solved by trial and error. They tested ideas in every possible way by combining inputs from different technical fields. The final solution they arrived at often came from an unexpected source. In order to find the solution, Canon had to create an environment where they could accumulate knowledge from different fields, exchange ideas, and test them. Thus, the co-location of engineers and researchers was the critical basis of this environment. Innovation took place in proximity specifically for those technical reasons.

Chapter 6
Innovation and Geography

This final chapter synthesizes findings from the preceding three chapters. First, I will review the geography of product development based on our framework of contemporary globalization. After summarizing why co-location was critical, I will propose the microdynamic theory of innovation with three engineering and technical features of generating innovations.

Then, the final section of the chapter will apply these findings to the role of information technology and the role of location in this era, with an intricate combination of being global and local. It will further extend discussions about how to conceptualize innovation and firms in the context of geography.

6.1 The Geography of Product Development

Using the framework of qualitative intensity and extensity of globalization, we confirmed that the majority of product development in the three cases was concentrated in the core region of each firm: the southern Tokyo region for Sony and Canon, and the Toyota-Nagoya region for Toyota. There was some long-distance coordination, but it was limited to peripheral functions.

Project members for Sony's Vaio were located at Shiangawa and Fujisawa only; the distance between these two points was 45 km. Almost all units for Toyota's Prius development were stationed at the headquarters and Development Center 2 in Toyota City. While an engine system group, BR-VF, was initially located at Higashi Fuji Technical Center, 200 km east of Toyota City, and these two places were connected with a major highway and frequent intra-firm helicopter traffic, they later found this two-hour journey a burden. Thus they moved BR-VF to the headquarters location during the second half of the project period.

During its long 13-year process of developing the Bubble Jet printer, Canon located its development teams in four consecutive places: a research laboratory at the university at Meguro (1977–1979), the Central Research Lab at Atsugi (1979–1983), the Technical Center at Hiratsuka (1983–1989), and a development center at Tamagawa, which was next to its headquarters in Shimomaruko (1989 to present). All these places are in the southern Tokyo metropolitan area, within a one-and-a-half-hour journey from the headquarters. More importantly, all the development-related people moved together from one place to another.

There were some development activities conducted outside each firm's core region, including overseas R&D laboratories. However, the role of these labs was highly limited. Sony mobilized its San Jose and Brussels labs only to translate the label and brochure for the Vaio. There was no role played by overseas research labs in Canon's case. On the other hand, Toyota used its overseas unit more extensively. The idea for the exterior design came from the Calty Design Center in Newport Beach, California. However, it must be noted that this exterior design process was not a central part of development. The scale of design, such as the overall length, width, and height, was predetermined by G21, the main development group at Toyota City. Once the scale was fixed, the exterior design had little engineering interface with the development of other components in terms of design and technological advancement.

These overseas activities were global, based on the framework of qualitative intensity and extensity. However, the global R&D connections had a hierarchical structure and highly limited functions. Essentially, these overseas activities interacted in a limited way with specific components of the product and did not involve major technological advancement. In other words, a change made at an overseas lab unit would not affect the whole system. This pattern of long-distance coordination had a hierarchical structure because key groups of project members in the core region were in charge of product definition, technological development, and overall coordination. They then assigned specific, limited tasks to overseas units.

6.2 Reasons for Co-location

This research found complementary findings to the three schools of thought discussed in Chapter 2. The local institutional link among the intraregional assets, as they are called by the agglomeration school, was critical. Suppliers provided key components: the new powerful batteries from Matsushita Batteries to Toyota and a smaller yet higher resolution display from Toshiba to Sony. A research university and a local fire department contributed to Canon's research. Engineers acquired ideas on how to solve problems through their social networks, though in these cases they were mostly networks within each firm. Tacit knowledge was definitely involved when project members discussed the stylishness of all three products.

Nonetheless, there were fundamentally different microdynamic processes of innovation that forced the location of development units in each firm. These were the engineering and technical aspects of innovation making. More specifically, there were three aspects: a product as a complex system, the interdisciplinary and serendipitous nature of technological development, and prototyping and testing. Next we will discuss each of the three in detail.

6.2.1 The Product as a Complex System

While a final product is often considered to be an innovation, this detailed microdynamic analysis showed that there were many *innovations* involved within a product. There was a complex coordination process between various components within a product. In the three cases, the number of components ranged from approximately 100 in the printer, to 1,100 in the laptop computer, and more than 30,000 in the automobile. Development of a new product meant the development of many new components, if not all of them.

A new product would need to function as a complete system, and each component had to be integrated within this system. This means that a change in one component could call for a modification to other components, as well as to the whole system. We discussed this in Toyota's Prius development, for instance. When they found a large fluctuation in battery levels and subsequent temperature changes, the engine system integration team had to request an upgrade of the cooling system. This resulted in structural changes to the car, such as where to locate batteries, how the cooling system would function, and even additional cooling equipment. In short, product development was a complex and dynamic process in which the firm had to adjust to an evolving system and situation.

Normally, there was one division in charge of the development of each specific component or its related parts. Coordination between components meant coordination among these various divisions. There was a constant flow of people between divisions to exchange ideas, provide feedback, and make modifications. Such coordination and induced changes became so frequent and crucial to the creation of the product as a system that each division involved had to be co-located. This co-location was within a one- to two-hour journey for each firm—it was not randomly selected anywhere around a metropolitan area. The technical centers and central research labs were established within such distance so that project members could meet when they needed to exchange ideas intensively and extensively with face-to-face communication, whenever they faced problems.

6.2.2 The Interdisciplinary Nature of Technological Development

Each firm introduced a new product with new features often through technological advancement. The three chapters revealed that an interdisciplinary approach often led to such technological advancement.

Making Sony's Vaio was a battle for miniaturization. After the removal of a fan, they had to establish effective aerodynamics, which was an integrated issue of both electrical and mechanical engineering. Additionally, the integration of new material with the aid of chemical engineering turned out to be effective and essential. Canon's Bubble Jet faced a series of technical difficulties over 13 years. For example, the problem of *cabitation*, a shock of bubble burst damaging the heater, was identified through the use of precision technology and a photo-recording technique. It was solved by inserting a filter between the nozzle and heater, involving the fields of chemistry and electric engineering.

Sometimes, such integration of technological fields happened serendipitously: Nobody predicted such a method, certainly not at the beginning of a project. Vaio's designer, Asawa, described one such unpredictable process. Even after he and his project leader had selected major components, some project members or other non-project colleagues later brought one lighter or smaller component after another, which necessitated the further redesign of prototypes. When Toyota's Sasaki was trying to establish a simulation algorithm for the hybrid system, he was saved by an unexpected visit by his former colleague, who suggested four simplified factors for determining driving conditions. Often, the knowledge necessary for innovation was somewhere within the firm, but not every person was necessarily aware of it or could readily access it. Some serendipitous incident would bring such knowledge in and lead to a technological advance.

With the three cases, which are conventionally categorized in the electronics and automobile industries, the commonly integrated fields were electric, mechanical, and simulation engineering, as well as physics and chemistry. Experts in all fields had to get involved in the process, and physical co-location was the most efficient way of achieving this. The role of each firm was to provide an environment where project members could exchange ideas flexibly and serendipitously.

6.2.3 Prototyping and Testing

Project members had to produce prototypes and test the functionality of the product. Additionally, engineers always had to test the interface of the product from a consumer's standpoint. When the Vaio team discussed laptop design, they had to produce a paper or foam polystyrene model. Even after they decided on the design, they produced prototypes three times to confirm that their concept had been successfully translated into the final product. Prototype testing by the Evaluation Group at Toyota was crucial not only to evaluate whether the car functioned, but also to make improvements. Although each component was developed in great detail, Toyota could detect system coordination problems only through testing a prototype. The Evaluation Group pointed out a physical shock and noise problem when the hybrid system was switching between the batteries and the gas engine. They provided this feedback to respective divisions, which necessitated further mechanical changes. Even though components functioned properly within a subsystem, testing the whole system was a different story. Canon's staff paid extra attention to incorporating as many human senses as possible: sight, touch, sound, and smell. The relevant point was how a consumer would think about this printer if it was sitting next to him or her, instead of how the system was technically organized as seen on a computer screen perceived by an engineer.

This prototyping and testing was also a process of sharing tacit knowledge. One of the critical tacit knowledge aspects for product development was what project members considered to be a good product. They did not simply believe the cheaper, the better. Each member had to have a subtle understanding of the uniqueness of the product. Lester and Piore (2004) call this "the interpretive process" of innovation making.

The Sony Vaio team came up with the basic concept of "portable and good-looking," but had a particular expectation of what was meant by "good-looking."

This understanding was part of their previous experience and was further developed through their discussions. Most of it was not written in meeting memos, but shared in their minds. Prototyping made this process more explicit: whether what one engineer considered "good-looking" was also "good-looking" to others. Masaki, Canon's research director, succinctly referred to this tacit dimension by saying: "If you don't know the Canon culture, you cannot make a Canon product" (2004, Chapter 5). What Masaki meant was that, to be effective, each project member within the organization needed to understand a whole set of expectations, including what Canon people would consider a good product, how they organize projects, how to exchange information, etc. This understanding and learning experience could easily take place between people who met frequently and shared various forms of information.

6.2.4 Unifying the Three Features of Innovation Making

All three of these features of innovation making forced each firm to co-locate their development resources and personnel. Co-location minimized the transaction costs of coordinating each complex system. Ideas for technological advancement came most often if people were co-located and exchanged their knowledge in an interdisciplinary and flexible way. Finally, prototyping and testing required the physical presence of the product and its evaluators.

Furthermore, the process of making innovation was often messy, lengthy, costly, and unpredictable. Creating a new product was a continuous process of trial and error, and it was not a process where one engineering genius designed everything from beginning to end. Instead, at the beginning of each project, nobody knew what the final product would be. They defined it only by working on it. The process was even more complex than "learning by doing" (Young 1991; Irwin and Klenow 1994) and could be described as learning by interactively responding to mistakes, again and again.

These findings about the engineering and technical features of innovation can provide new insights that previous studies could not. Both the social network and tacit knowledge schools emphasized the importance of face-to-face communication. However, they did not explain exactly what face-to-face communication was needed for, except to promote trust or to share tacit knowledge. This study can connect the need for face-to-face communication and the technical evolution of innovation in the daily operation.

Additionally, one question remained: Once people meet and share the same organizational knowledge, can they exchange and generate new knowledge over distance? The best guess from these schools was yes. In fact, in her recent analysis of transnational networks, Saxenian and her colleagues (2002, 2006) argued that Indians, Chinese, and Israelis working in Silicon Valley were able to bring Silicon Valley's distinctive institutional practices back to their home countries. Nonaka and Takeuchi (2000, 297–333) argued that in the cases of Nissan's Primera model and a joint project by Mitsubishi (Japan) and Caterpillar (U.S.), they shared tacit knowledge on a worldwide scale as long as the firm transferred its engineers.

Rather than argue over whose finding is right or wrong, the more important question is to deconstruct these broad concepts and investigate which aspects of the institutions or their tacit knowledge were transferred. The institutional aspect observed in Saxenian's or other transnational literature was limited to the frequency and extensity of networks, and the creation of venture capital. The analysis of tacit knowledge can be controversial because specification of tacit knowledge is, by definition, difficult. As a result, previous studies tended to discuss the scope and role of tacit knowledge quite broadly and vaguely. They assumed that some kind of knowledge was exchanged, but did not know exactly what it was.

In contrast, this book calls for more cautious analysis, because the engineering and technical aspects of generating innovation are difficult to move even though people may move. Therefore, a critical exercise is to reanalyze the feasibility of global R&D coordination by further considering the microdynamic dimension discussed in the book. Thus, some institutions and tacit knowledge may be transferred relatively easily if they do not involve the engineering and technical dimensions of innovation making. Because institutions and tacit knowledge are broad concepts, generalizing from a few findings is not the most constructive analytical process.

6.3 Further Discussions and Implications

Following the synthesis of findings so far, this last section of the chapter will extend the discussion to the following four topics: the role of information technology, the role of location, innovation in economic geography, and firms in the regional economy.

6.3.1 The Role of Information Technology

In each instance discussed, information technology facilitated the product development process, but did not substitute for co-location. Firms applied information technology to complement the advantages of proximity. In this discussion, we need to differentiate between two subcategories within what is broadly termed "information technology": communication technology and information technology proper.

First, communication technology is a system for transmitting information between two or more actors. This includes email, the Internet, and intrafirm blogs. Although little has been mentioned so far about the use of the Internet during the product development process, email was widely used to exchange information. This includes use for logistical purposes, such as setting up meeting times and locations, and for follow-up purposes after meetings. All the firms actively used an intra-firm blog to exchange information on a wider scale. The Vaio's project members posted major debates about the product concept after each meeting, enabling recently-added members, such as the designer Goto, to catch up with the rest of the team quickly.

Early on, Toyota circulated memos on paper among project members regarding the functional information of specific components, predominantly their limitations, such as durability and the way components broke. This was useful, but it took several days to receive the most updated information. With the electronic project blog, distribution of this information was easier and faster. Sometimes one person posted problems, and others made suggestions for improvement on an hourly basis. Although the Prius was officially a clandestine project, the blog was accessible to many people inside the company. Occasionally people outside the project but with an interest in the Prius provided useful suggestions on technical matters (Yaegashi, interview, December 4, 2004).

On the other hand, none of the cases provided evidence that the use of communication technology replaced face-to-face meetings. Whenever major technological problems emerged, there was always a meeting of key people to discuss the matter. They used communication technology to complement the dense information exchange of face-to-face meetings.

The second type of technology is information technology, a system for reprocessing input for a specific, technical function. This includes computer-aided design (CAD), computer-aided engineering (CAE), and simulation software. With the introduction of CAD, component redesign at Toyota decreased substantially from 10,000 to 500 changes per model (Hino 2002, 314).[1] Sony created a system to simulate aerodynamics to control the temperature inside its laptop, and Canon did likewise to simulate hydrodynamics to estimate the shape and timing of ink droplets.

Nonetheless, information technology itself did not eliminate all problems. There were some aspects that current information technology could not deal with. The Vaio project's mechanical engineer achieved a higher density of components in the laptop by intentionally breaking the conventional CAD rules. Under CAD, each component was, by industry default, designed with some extra buffer space so as to not collide with adjacent components. A mechanical engineer considered how much buffer space was actually needed depending on the functions of adjacent components, and further calculated how many millimeters could be reduced case by case. Here, human flexibility and creativity superseded the computer's standard calculation.

In each case, they created prototypes over several stages. As described earlier, making prototypes was critical not only to test the physical functionality of the product, but also to analyze the interface and to improve it. Some tools in current information technology were able to test whether the product actually worked, e.g. electricity could reach the necessary components, etc. However, information

---

1   However, there could be factors other than CAD that contributed to this reduction, such as changes in organizational structure to enhance better coordination.

technology was ill equipped to test the interface aspects, such as the feel, sound, and visual appeal of the product to consumers. Humans always tested the product from every possible aspect, analyzed the tests, and suggested improvements.

Information technology did not eliminate all the problems in the development process. While the need for redesign of an automobile in development was substantially reduced, there were still 500 readjustments. This plainly meant that the whole development team still had to deal with 500 redesigns, not a small number at all. Additionally, the use of information technology produced new kinds of problems. In the Prius case, there was a major mistake in the use of two-dimensional CAD, and a group of staff members rushed to the factory to come up with an adjustment. The same kind of error can happen with more recent three-dimensional CAD. Some human errors were unavoidable, and co-located humans had to fix them quickly. Additionally, Toyota's cooling system team realized that there was an error in the temperature in a simulated model of the cooling system due to a seasonal difference in the outside temperature and to the higher functionality of components. Simulation programs were, after all, simulations, i.e., simplified models that could not incorporate everything in reality. No matter how well the project was planned and information technology advanced, there would be some errors and insufficient calculation that would require readjustment. Given the complexity of the modern consumer product and the necessity to coordinate among components, the objective of the project teams was to minimize the adjustment periods. Information technology was one of the tools used to achieve this, but only one of the necessary tools. Co-location was the most effective solution.

Advocates of "death of distance" theories or other optimistic globalists (O'Brien 1992; Caincross 2001; Friedman 2005) emphasized the simple possibility that IT may replace a specific function. They tended to ignore one of the most critical factors in the economy: market competition. Incorporating this market pressure and the need to develop the product in a timely manner changes the dynamics of the calculations made by each firm. For instance, Canon was and still is competing against a different type of ink jet printer produced by HP, and Toyota against hybrid cars produced by Honda and other automobile firms. Therefore, the earlier a firm could introduce a new product, the more profit they could gain. Not to mention a reputation for pioneering in a particular market. Additionally, firms always had an incentive to shorten the development period because that could save on costs. As a result, each firm was keen to minimize the development period. Thus, whether information technology could create the virtual reality to allow operation from remote places was not exactly in each firm's best interest. The critical question was how they could use information technology to become even more efficient and effective. The role of information technology was to facilitate the process, but not to replace the advantages of co-location.

6.3.2 The Role of Location

The Power of Being Local To Go Global

My research observed how each firm organized personnel, knowledge, and resources at a specific locality, and highlighted the complexity of the innovation and production system. It further provides insights into the role of location. Before discussing conclusions from the research, we should review five fundamental principles that this book has been investigating with its theme of globalization and innovation. First, competition takes place on a global scale. Firms try to sell as many products in as many places as they can. Second, as a result, there is continuously increasing pressure to innovate. Making a good product is not enough, because competitors can come up with similar competing products. Third, the creation of innovations mostly comes from interaction on the local level, as we have seen. Thus, fourth, multinationals operate by developing products at the local level, while they sell worldwide. Lastly, information technology is critical for the global coordination of manufacturing, logistics, and sales activities, and it does facilitate product development. However, it does not substitute for the advantages of being co-located, particularly where product development is concerned.

In this sense, globalization has not penetrated every corporate activity, but only specific functions. Multinational firms operate in a duality: global manufacturing and sales activities and local innovation systems. In the meantime, multinationals often claim that they operate R&D centers around the world and integrate the best talent and knowledge to create innovations. However, the case study from this microdynamic approach demonstrates that the reality is not that simple. It is a messy, unpredictable process lasting several years with numerous trials and errors. Distance becomes an obstacle for firms.

In this age of competition and globalization, with its dual geographic nodes of being local and global, the role of location has not diminished. In fact, more market competition can highlight the importance of being local to innovate. In the three cases here, the scale of locality was a metropolitan area. The two-hour distance was a geographic scale at which people could meet and exchange knowledge on a daily basis, whenever necessary. They could most effectively coordinate the long, complex, unpredictable development process within this distance.

In the past few years, there has been major concern about the "hollowing-out effect" in both the U.S. and Japan.[2] The essential debate was around the moving of many jobs, investment, and technologies to emerging countries, most notably to mainland China and India. The most recent debate included the outsourcing of white-collar jobs and even R&D activities, and it was claimed that the enormous size and high-quality labor of these two countries made cheap outsourcing of even R&D possible.

---

2   See, for example, debates of the National Academy of Science (Macher and Mowery 2008) for the U.S., and the Ministry of Finance (2002) and Cabinet Office (2002) for Japan.

The research in this book suggests that such concern was exaggerated, at least where R&D activities are concerned. First, the creation of innovations is not a type of activity in which one side tells the other what to do, and the other simply executes the command, as outsourcing typically constitutes. If what to do is already defined, there naturally is no innovation. Instead, people and firms involved in innovation have to figure out what they will do by going through it, which creates the lengthy, complex, and unpredictable development process and the need to coordinate it. Second, the need for co-location tells us that, unless the entire development center is moved to one place, it will not function effectively. Alternatively, a small research unit with limited functions may be easier to relocate; however, by definition, this will have little spillover effect in the local economy and a limited impact on the home country. Furthermore, the thick linkage among local agglomeration factors means the need to introduce suppliers, universities, and other players into the mix. Relocation of the entire set of economic actors is extremely difficult, if not impossible.

At a different level, the current local economic development policy based on the cluster theory (Porter 1998b, 2000) focuses on upgrading to value-added activities, such as R&D, or attracting high-tech sectors. With the same logic, this is hardly a promising strategy. It is almost impossible for government to create all the economic actors from scratch.

On the other hand, the strength of local connections indicates the need for enhancing the capacity to create innovation that already exists in the area. Policy makers should not blindly target R&D or high-tech industries, but assess the existing R&D capability in their region and the extent of such a network. In some cases, it will be more promising to promote what already exists than to shift to sexy high-tech or R&D activities.

Some Differences in Co-location

This was not a comparative study, in which I would attempt to explain how the differences in input have resulted in the differences in outcomes. That type of study would necessarily have to find differences in the outcomes, i.e., the geographic distance in product development. Instead, this was an exploratory case study to investigate the linkage between the process of generating innovation and geography, and the purpose was not to estimate how far apart development units may be located. Therefore, finding a similar outcome in all three cases, the two-hour proximity, was not a problem. It simply confirmed that the finding can apply to more than one firm and one industry.

On the other hand, there were some, though minor, differences in the distance between development and production units in each case. What conclusion can we draw from that? What firm or industrial factor might have caused it?

The distance between development units constituted a range, from almost zero for Canon, to 45 km for Sony, and 200 km for Toyota. There was a similar range for the distance between the headquarters of each firm and its development units: 1–52 km for Canon, and the same 45 km and 200 km for Sony and Toyota, respectively. With this data, Toyota's case seems to involve the most distance. However, this was not due to any specificity in the automotive industry. Toyota established the Higashi Fuji Technical Center in 1966 (Toyota 2005a) to test-drive prototypes, which requires a vast amount of open space. Since they could not find available land in the Toyota-Nagoya area, which was already urbanized by that time, they located it in an unurbanized site, Higashi Fuji. Over the years, they expanded its functions beyond test drives and evaluation. When a firm starts a new product development project, it uses its existing facilities. It usually does not establish a new R&D site for each product. Therefore, the variance in distance reflected each firm's history rather than industrial differences. All the research could suggest was that the two-hour distance was important, and that differences beyond that were not critical.

What mattered was the location of production sites once products were developed. Sony EMCS, the subsidiary for production of the Vaio, was located in Nagano, a three-hour journey, and Canon's factory was in Ami, Ibaragi, also a three-hour journey. On the other hand, all four of Toyota's factories were located in the core Toyota City. Why did Toyota need all its factories so close?

The automobile industry requires a vast number of components (30,000 per vehicle), which are heavy and bulky. Transportation and inventory of the components and the final product require complex and costly coordination, and firms minimize these costs by operating in proximity. Chapter 4 discussed the just-in-time *kanban* system developed by Toyota and how it could affect the cost of operation. Additionally, the automobile is an integrated product. Each final assembler has different specifications, along with quality and durability requirements for each component. In other words, an engine or brakes for a Toyota car can be used only for specific models and cannot be applied to Honda's or Isuzu's cars. This forces its suppliers to produce specialized components for each automobile assembler. This is in contrast to the modularity of the computer and electronics industry, in which components are standardized and interchangeable (Sturgeon 1997; Ernst 1997a; Hagerson and Eisenhardt 2000). In this modular computer industry, for instance, the central processing unit (CPU) of a PC can be Intel's or AMD's. These industrial differences caused the variation in factory location. For Toyota, it was most efficient to carry out both prototyping and mass production at a core location. With products of less complexity and bulkiness, Sony and Canon could operate mass production at a distance, though not too far away.

6.3.3 Innovation in Economic Geography

One of the major aims of this book was to open the black box of innovation and explore the mechanisms that create it. While the box is deep and wide, and many areas still remain black, we have successfully uncovered some parts of it. This section will extend the discussion with two more aspects: innovation as a process and its multidimensional evolution.

Innovation as a Process

Previous research in economic geography and most business administration treated *innovation* in a singular form. In this way, the studies treated innovation vaguely and holistically, and did not consider the heterogeneity of innovations and the different mechanisms employed to create them. Additionally, innovation was something resulting from competition and collaboration among firms, suppliers, customers, and other institutional factors. In this sense, the literature treated innovation only as an outcome. However, the complexity of the products and technology in this study revealed that a firm has to accumulate many *innovations* to create a new product. There were many intermediate technologies that each firm had to create to use as inputs. Then, those new technologies were integrated to generate the final product. Thus, innovations were not only an outcome, but also a process.

Moreover, the development of innovation was shown to be not a positive or hierarchical process, but an organic one. Most theories in the social sciences are prone to rationalistic and positivist thinking, which generally assume that the designing of a new product was the end of an innovation process and that production of it would subsequently follow. This rationalistic thinking was not limited to the market rationality in neoclassical economics, but carried as well to the field of management of technology broadly influenced by Frederick Taylor (1903, 1911). This scientific management concept was widely applied to the Fordist management style in the early half of the twentieth century and continued to total quality management in the 1980s and the Sig Sigma movement in the 1990s. It essentially assumed that the most efficient method of production could be scientifically calculated and then applied in practice.

In contrast, the lengthy and unpredictable aspect of innovation means that such scientific calculation does not shape the final product in reality. Innovation is not driven by a scientific method or by a single genius engineer. Engineers and managers investigate a number of possibilities to see what could work better or what would work at all. Therefore, this was not a hierarchical process with a top-down approach, because trial and error was conducted at each component level and eventually integrated as a system, in a bottom-up approach. At the same time, the chief engineer of each project had to determine the final configuration of each component and integrate the product into a system. Therefore, this requires a mixture of bottom-up and top-down approaches.

This non-hierarchical process of innovation was further confirmed by a variety of paths in the stages of innovations. Chapters 3–5 followed a common structure of four parts: (a) the search for technology, (b) concept development, (c) component development, and (d) the preparation for mass production. In actual practice, the development phases varied in some ways from this pattern.

Table 6.1   Summary of development phases

#	Sony	Toyota	Canon
1	Concept Development	Concept Development	Search for Technology
2	Component Development	Search for Technology	Concept Development
3	Preparation for Production	Component Development	Component Development
4	Mass Production	Preparation for Production	Preparation for Production

Source: Author.

Table 6.1 shows that each case had a unique beginning and a set of development phases, so there was no universal path for innovation making. Development of a new product could either start with a concept or with new technology. While technology could determine functions, size, and some features of a product, market-based product concepts also had to match available technology. Of the three cases, it was only Canon's Bubble Jet that started with the technology defining the product. Sony's and Toyota's projects started with a product concept, then matched technology to the concept. Thus, contrary to the hierarchical theory that assumed a single path trajectory for innnovations, technology itself did not necessarily determine the final product. The lengthy, unpredictable process of innovation making was tightly linked to its organic and non-hierarchical nature.

The Multi-Dimensional Evolution of Innovation
The previous section outlined the interdisciplinary nature of technological development. This nature leads to another critical feature of innovation, its multidimensional aspect.

Innovations identified in the previous literature had four forms: a new product, a new method of production, a new organization, or a new market (Schumpeter 1934, 66; Freeman 1982, 7). This study used only the new product as its ultimate observational unit, yet the three case studies revealed evolution in other types of innovation during the process: a new method of production, a new organization, and even a new market. For example, to mass-produce the Prius, Toyota's Takaoka Factory underwent major rearrangement in the *method of production* to establish a separate assembly line, because of the lower production volume. Production engineers, factory managers, and staff from G21 conducted a brainstorming exercise to create the new method of production. The establishment of G21 was the creation of a new *organization* for Toyota. Conventionally, one of the development centers would have been in charge of conceptualizing the new car and coordinating the overall product. Detached from any development centers and directly under the headquarters unit, this unique unit, G21, played a pivotal role in creating a car with new technology and new concepts.

The Sony Vaio created a new market in non-office use, portable and stylish laptops, and Canon's Bubble Jet similarly initiated a market for home-use printers.

The high efficiency and environmental friendliness of the Prius started the race in a completely new *market* for hybrid cars, with competition between Honda which introduced the Insight in November 1999 and GM, which finally started to introduce hybrid models in 2008.

Thus, there was high interaction between the four forms of innovations. Creation in one form of innovation triggered the creation of other forms. Innovations were multi-dimensionally connected and evolving. Each firm was a master at managing this accumulation in the number and various forms of innovations.

6.3.4 The Firm in the Regional Economy

Economic geography or regional economy,[3] by definition, is the study of economic activities. It is ironic that this field has paid little attention to the role of a firm even though a firm is the fundamental unit in the economy. This is a crucial limitation because it is analytically dangerous to advance theories of economy without knowing the intentions or objectives of the fundamental actor. Based on the case findings, this section will analyze what a firm means in the regional economy.

Strategic Intention of a Firm
The field of economic geography has traditionally established little connection with theories explaining what actually constitutes a firm or why it exists (Taylor and Asheim 2001, 315).[4] Instead, it treated a firm almost as a "phenotype" (Noteboom 1999) or "phylogeneric ontology" (Maskell 2001, 331), something that exists in the economy but without a shape or an intention. At best, it treated a firm as an entity intended simply to maximize profits, an assumption borrowed from neoclassical economics. This is mostly because the field has focused on analyzing populations or aggregate units, such as industries or regions, rather than individual units.

Thus, the theory defined the firm as little more than one of several economic actors. Additionally, the theory considered the creation of innovation only in neutral terms. In other words, the theory suggested that if you bring firms, suppliers, and other agglomeration factors into a box called a region, and shake it well, some kind of innovation should come out.

In reality, a simple collection of economic actors is not sufficient to create innovations. Ideas do not sit in a region, but must be possessed, organized, and applied by a specific actor. Technology itself does not produce economic values, define markets, or appeal to consumers. Someone has to apply it. Someone has

---

3   Economic geography is a subfield in geography and regional economy is a subfield in city or urban planning. They are virtually the same, and I treat them synonymously in this book.

4   Exceptions do exist; for example, Hodder and Lee (1974), Storper and Walker (1989), Amin and Thrift (1994), Lee and Wills (1997), Schoenberger (1997), Cooke and Morgan (1998), and Amin and Cohendet (1999).

to define the scope of a product. There must be a strategy and an intention by a specific actor to organize corporate and agglomeration resources. It is this strategic role that only a firm (or perhaps also an individual) can play in the economy so that innovations take on concrete, commercial form.

In this sense, technologies are ingredients, and a firm is a strategic chef who creates the recipes and menu. The presence of high-quality, fresh ingredients is insufficient to make a good dish; simply mixing them collaboratively is still insufficient. The chef selects the styles and flavors of the dish, as well as which ingredients to use and how they are to be mixed. Without a chef, these ingredients never become a great meal.[5] Therefore, a firm is not just one of several economic actors, but the central unit that defines, creates, coordinates, and applies innovations to commercialization. A firm shapes and frames the boundary of innovations.

Internalization of Agglomeration Factors
The three cases presented show how innovations were made through the interactions between various economic actors: the lead firm, suppliers, universities, and other local institutions. On the other hand, the cases did not actively identify the involvement of other economic actors typically emphasized in the agglomeration literature: the labor market, core customers, and investing institutions because the firms internalized them. The labor market was not externally open, but rather functioned internally—within each firm. Engineers with specific skills were transferred from one division to another in response to needs. The firms studied were large, multinational corporations with plentiful capital resources. An investing institution, such as venture capital, was not needed to support the development projects financially. Instead, corporate executives, sometimes instantaneously, approved the procurement of specific components and machinery. This procurement could easily involve hundreds of millions of dollars of support over years.

There was little evidence that customers provided feedback. This type of feedback between the market and the firm usually happened after one product was released and before another product was designed, not within one cycle of product development. However, each firm simulated this feedback mechanism by incorporating marketing perspectives during the development process. Marketing personnel and product evaluation divisions played a significant role here.

These internalization efforts provide an additional connection between the theory of economic geography and the theory of a firm. The latter, called institutional economics, tried to answer the question of why a firm exists by analyzing how it determines whether to produce goods or services internally or to purchase from the market. The insightful analysis of Ronald Coase (1937; Coase et al. 1991) proposed that firms exist to reduce the transaction costs that could not efficiently be overcome in market exchange. While institutional economists normally limit the notion of transactions of goods, services, or information through the market,

---

5   I owe this analogy to Beth Chapple.

this book revealed that firms play even more extended roles. Thus, a firm exists not only to internalize functions of the market, but also to internalize the functions of agglomeration factors within a region. At the same time, this does not necessarily mean that the more internalization takes place, the better. Just as a good balance between purchase on the market and internal production is critical, there must be a balance between how much to internalize and how much to procure from other agglomeration factors. The firm is the central and strategic unit for calculating this balance.

6.4 A Final Remark

What did this study do? Through an in-depth study of product development in three cases, this research examined the connection between the process of generating innovation and geography. Additionally, it provided an analysis of connections among cities around the globe, though only scarce and secondary connections were observed. We have consistently used the framework of the intensive and extensive connections in the context of contemporary globalization.

This was a study of innovation and globalization, but without necessarily having a region as the central unit of analysis. Observations can come from analyzing specific actors, such as firms, and activities, such as product development. Due to its phylogeneric ontology, the theory of economic geography has traditionally looked less at specific firms or activities, which has led to theoretical limitations especially when analyzing the process of creating innovation and globalization. Markusen (1994) called for "studying regions by studying firms." Moreover, we can uncover the process of creating innovation by analyzing not only firms, but also activities. Future research should explore connections among actors, activities, and geography.

List of Interviewees

Chapter 3: Sony's Vaio

Aoki, Masayoshi. General manager. Corporate Headquarters. Sony. September 21, 2004.

Endo, Eishi, Ph.D. Assistant manager. Energy R&D Department. R&D Division. Energy Company, Sony. May 7, 2004.

Goto, Teiyu. Designer. Creative Center. Sony. September 22, 2004.

Iguchi, Akira. Software leader. IT Company. Sony. December 4, 2004.

Ito, Susumu. Senior General Manager. VAIO Design Department. VAIO Design Division. Sony EMCS Corporation; Deputy General Manager. Departmnet No.2. IT Company. IT & Mobile Solutions network Company. Sony. September 21, 2004.

Katsumoto, Shuzo. Section Chief. Center for Environment Promotion. Sony. August 10, 2003; August 6, 2009.

Kurosawa, Nozomi. Assistant Manager. International Public Affairs. External Relations Department. Sony. July 21, 2004.

Miyano, Akihiro. Electrical designer. IT Company. Sony. December 4, 2004.

Sakaguchi, Kei. General Manager. Corporate PR Department. Corporate Communications. Sony. July 21, 2004.

Sanakata, Keita. Assistant Manager. Corporate PR Department. Corporate Communications. Sony. July 21, 2004.

Tambata, Ippei. Designer. HIGP. Design Center. Sony. September 24, 2004.

Toyoda, Mami. Human Resources Department 1. Sony. September 20, 2003.

Yokote, Hitomi. Senior Manager. International Public Affairs. External Relations Department. Sony. July 21, 2004.

Chapter 4: Toyota

Hoshino, Ryuzo. Professor. Tokyo Design and Art University. April 24, 2004.

Ikoma, Munahisa. Director. Panasonic EV Energy. Toyota Corporation. September 28, 2004.

Kaneda, Shinji. Manager. Body Technology Division. Takaoka Factory. Toyota Corporation. September 29, 2004.

Kitabayashi, Nahoko. Project General Manager. Global Economy and Industry Analysis Group Research Division. July 15, 2004.

Kitagawa, Fumikazu. Senior Consultant. Strategic Business Department. Nomura Research Institute. June 25, 2004.

Kotani, Takeshi. Engineer. Engine Technology Division 3. Toyota Corporation. September 29, 2004.

Kurihara, Ikuo. Project General Manager. Research Division. Toyota. July 15, 2004.

Kusajima, Takayuki. Assistant Manager. Global Economy and Industry Analysis Group. Research Division. Toyota. July 15, 2004.

Matsui, Hideaki. Engineer. Drive Train Technology Division and Business Reform Vehicle Fuel Efficiency. Toyota Corporation. September 28, 2004.

Muramatsu, Motoyasu. Chief Evaluator. Product Evaluation Team. Development Center Division 2. Toyota Corporation. September 28, 2004.

Nagai, Tsuneo. Senior Consultant. Information Technology Industry Department. Nomura Research Institute. June 15, 2004.

Suzuki, Kenji. General Manager. Research Division. Toyota. July 15, 2004.

Yaegashi, Takehisa. General Manager. BR-VF. Toyota. Torrance, CA. December 4, 2004.

Chapter 5: Canon

Omata, Satoshi. Senior General Manger. Leading-Edge Planning & Administration Center. Leading Edge Technology Development Headquarters. Canon. September 24, 2004.

Taniishi, Shinnosuke. Staff manager. Development Process Planning Department. Core Technology Development Headquarters. Canon. September 24, 2004.

Hara, Shoichiro. Senior Consultant. Industrial Consulting Department. Nomura Research Institute. May 4, 2003.

Hajimoto. Independent Consultant. Former Director at Canon Central Research Lab. July 20, 2004.

Inamori, Kenichi. Managing Director. Canon USA Research. October 7, 2004.

References

Abramovitz, M. 1956. Resource and Output Trends in the United States since 1870. *American Economic Review* 46(2): 5–23.

AEA (American Electronics Association). 2003. Defining the High-Tech Industry: AEA's New NAICS-based Industry Definition. Washington DC: AEA.

Allen, T.J., and O. Hauptman. 1990. Substitution of Communication Technologies for Organizational Structure in Research and Development. In *Organizations and communication technology*, edited by J. Fulk and C.W. Steinfield. Newbury Park, Calif.: Sage Publications, 328.

Amin, Ash, and P. Cohendet. 1999. Learning and Adaptation in Decentralized Business Networks. *Environment and Planning D: Society and Space* 17: 87–104.

Amin, Ash, and N.J. Thrift. 1994. *Globalization, Institutions, and Regional Development in Europe*. Oxford: Oxford University Press.

Aoki, Masahiko. 1988. *Information, Incentives, and Bargaining in the Japanese Economy*. New York: Cambridge University Press.

———. 1994. The Japanese Firm as a System of Attributes: A Survey and Research Agenda. In *The Japanese Firm: The Sources of Competitive Strength*, edited by M. Aoki and R.P. Dore. Oxford: Oxford University Press, 11–40.

Aoki, Masahiko, and Ronald Philip Dore. 1994. *The Japanese Firm: The Sources of Competitive Strength*. New York: Oxford University Press.

Arimura, Sadanori. 1999. How Matsushita Electric and Sony Manage Global R&D. *Research Technology Management* 42(2): 41–52.

Asheim, Bjorn. 1998. Territoriality and Economics: On the Substantial Contribution of Economic Geography. In *The Swedish Geographical Yearbook*, edited by O. Jonsson and L.O. Olander. Lund, 98–109.

Asheim, Bjorn, and Meric S. Gertler. 2005. The Geography of Innovation: Regional Innovation Systems. In *The Oxford Handbook of Innovation*, edited by J. Fagerberg, D.C. Mowery and R.R. Nelson. Oxford: Oxford University Press, 291–317.

Audretsch, David B., and Maryann P. Feldman. 1996. R&D Spillovers and the Geography of Innovation and Production. *American Economic Review* 86(3): 630–40.

Bartlett, Christopher A., and Sumantra Ghoshal. 1987. Managing Across Borders: New Organizational Responses. *Sloan Management Review* (Fall): 43–53.

———. 1993. Beyond the M-Form: Toward a Managerial Theory of the Firm. *Strategic Management Journal* 14: 23–46.

———. 1998. *Managing Across Borders: The Transnational Solution*. 2nd ed. Boston, Mass: Harvard Business School Press.

Bateson, Gregory. 2000. *Steps Toward an Ecology of Mind*. Chicago: University of Chicago Press.

Bauman, Zygmunt. 1998. *Globalization: the Human Consequences, European perspectives*. New York: Columbia University Press.

Beaverstock, J., P. Taylor, and R.G. Smith. 1999. A Roster of World Cities. *Cities* 16(6): 445–58.

Beaverstock, J.V. 2005. World City Networks "From Below": International Mobility and Inter-City Relations in the Global Investment Banking Industry. In *GaWC Research Bulletin*. Leicestershire, U.K.: Globalization and World Cities.

Beaverstock, J.V., M.A. Doel, P.J. Hubbard, and P.J. Taylor. 2002. Attending to the world: competition, cooperation and connectivity in the World City network. *Global Network* 2(2): 111–32.

Beaverstock, J.V., R.G. Smith, P.J. Taylor, D.R.F. Walker, and H. Lorimer. 2000. Globalization and World Cities: Some Measurement Methodologies. *Applied Geography* 20(1): 43–63.

Belderbos, René. 2001. Overseas Innovations by Japanese Firms: An Analysis of Patent and Subsidiary Data. *Research Policy* 30: 313–32.

Bhagwati, Jagdish N. 2004. *In Defense of Globalization*. New York: Oxford University Press.

Bickman, Leonard. 2000. *Donald Campbell's Legacy*. 2 vols. Thousand Oaks, Calif.: Sage Publications.

Bordo, Michael D., Alan M. Taylor, and Jeffrey G. Williamson, eds. 2003. *Globalization in Historical Perspective, National Bureau of Economic Research conference report*. Chicago: University of Chicago Press.

Bordo, Michael D., Barry Eichengreen, and Douglas A. Irwin. 1999. Is Globalization Today Really Different than Globalization a Hundred Years Ago? In *NBER Working Paper*. Cambridge, MA: National Bureau of Economic Research.

Bourdieu, Pierre, and Loic Wacquant. 1999. On the Cunning of Imperialist Reason. *Theory, Culture, and Society* 16(1): 41–58.

Boutellier, Roman. 2008. *Managing Global Innovation: Uncovering the Secrets of Future Competitiveness*. 3rd ed. New York: Springer.

Brotchie, J.F. 1995. *Cities in Competition: Productive and Sustainable Cities for the 21st Century*. Melbourne: Longman Australia.

Brown, John Seely, and Paul Duguid. 2000. Mysteries of the Region: Knowledge Dynamics in Silicon Valley. In *The Silicon Valley: A Habitat for Innovation and Entrepreneurship*, edited by C.-M. Lee, W. Miller, H. Rowen and M. Hancock. Stanford: Stanford University Press.

Brusco, Sebastiano. 1982. The Emilian Model: Productive Decentralisation and Social Integration. *Cambridge Journal of Economics* 6(2): 167–84.

Cabinet Office. 2002. Annual Report on Economy and Finance. Tokyo.

Cable, Vincent. 1995. The Diminished Nation-State: A Study in the Loss of Economic Power. *Daedalus* 124(2): 22–53.

Cairncross, Frances. 1997. *The Death of Distance: How the Communications Revolution will Change our Lives*. Boston, Mass.: Harvard Business School Press.

———. 2001. *The Death of Distance: How the Communications Revolution is Changing our Lives*. Boston: Harvard Business School Press.

———. 2002. *The Company of the Future: How the Communications Revolution is Changing Management*. Boston, Mass.: Harvard Business School Press.

Camarota, Steven A., and Mark Krikorian. 2008. *Illegal Immigrant Population Dropping: New Report Finds Significant Decline Since Last Summer*. Washington DC: Center for Immigration Studies.

Campbell, Donald Thomas, Julian C. Stanley, and N.L. Gage. 1966. *Experimental and Quasi-experimental Designs for Research*. Chicago: R. McNally.

Canon, Inc. 2004a. Annual Report. Tokyo, Japan.

———. 2004b. Production Related Companies 2004 [cited October 14, 2004]. Available from http://www.canon.co.jp/about/group/list04.html.

———. 2005a. Canon Today 2005 [cited November 13, 2005]. Available from http://web.canon.jp/about/activities/index.html.

———. 2005b. Canon's milestone 2005 [cited November 11, 2005]. Available from http://web.canon.jp/about/history/main09.html.

———. 2006a. Canon Technology Inkjet Printer 2006 [cited January 9, 2006]. Available from http://www.canon.com/technology/ij/index.html.

———. 2006b. Canon Technology Laser Beam Printer 2006 [cited January 9, 2006]. Available from http://www.canon.com/technology/lbp/.

———. 2006c. Canon's Organization 2006 [cited January 14, 2006]. Available from http://web.canon.jp/corp/organization.html.

———. 2006d. Pixus Non-Durables: BJ-10v 2006 [cited March 11, 2006]. Available from http://cweb.canon.jp/pixus/supply/product/bj-10v.html.

———. 2006e. Production Related Technologies 2006 [cited January 6, 2006]. Available from http://web.canon.jp/technology/ij/index.html.

———. 2006f. Research & Development. Canon Inc. 2006 [cited April 20, 2006]. Available from http://web.canon.jp/about/r_d/index.html.

Cantwell, John A. 1995. The Globalisation of Technology: What Remains of the Product Cycle Model? *Cambridge Journal of Economics* 19: 155–74.

Carnoy, Martin. 2000. *Sustaining the New Economy: Work, Family, and Community in the Information Age*. New York: Russell Sage Foundation; Harvard University Press.

Carter, Susan B., and Richard Sutch. 1999. Historical Perspectives on the Economic Consequences of Immigration into the United States. In *The Handbook of International Migration: The American Experience*, edited by C. Hirschman, P. Kasinitz and J. DeWing. New York: Russel Sage Foundation.

Carter, Anne P. 2007. Measurement of the Clustering and Disperson of Innovation. Edited by K.R. Polenske, *Economic Geography of Innovation*. Cambridge: Cambridge University Press, 13–29.

Castells, Manuel. 1989. *Informational City: Information Technology, Economic Restructuring, and the Urban-Regional Process*. Oxford: Blackwell.

———. 1996. *Rise of the Network Society*. Information age; 1. Cambridge Mass.: Blackwell Publishers.

———. 2000. *The Rise of the Network Society*. 2nd ed. Oxford: Blackwell Publishers.

Census. 2000. Census of the United States. Washington DC: Department of Census.

Chandler, Alfred Dupont. 1962. *Strategy and Structure: Chapters in the History of the Industrial Enterprise*. Cambridge, MA: M.I.T. Press.

———. 1977. *The Visible Hand: The Managerial Revolution in American Business*. Cambridge, Mass.: Belknap Press.

Chapple, Karen, Ann Roell Markusen, Greg Schrock, Daisaku Yamamoto, and Pingkang Yu. 2004. Gauging Metropolitan "High-Tech" and "I-Tech" Activity. *Economic Development Quarterly* 18(1): 10–29.

Chinitz, Benjamin. 1961. Contrasts in Agglomeration: New York and Pittsburgh. *American Economic Review* 51(2): 279–89.

———. 1965. *City and Suburb; The Economics of Metropolitan Growth, Modern Economic Issues*. Englewood Cliffs, N.J.: Prentice-Hall.

Chiswick, Barry R., and Timothy J. Hatton. 2003. International Migration and the Integration of Labor Markets. In *Globalization in Historical Perspective*, edited by M.D. Bordo, A.M. Taylor and J.G. Williamson, Chicago: University of Chicago Press, 65–120.

Clark, Kim B., and Takahiro Fujimoto. 1991. *Product Development Performance: Strategy, Organization, and Management in the World Auto Industry*. Boston, Mass.: Harvard Business School Press.

Clark, Kim B., and Steven C. Wheelwright. 1994. *The Product Development Challenge: Competing through Speed, Quality, and Creativity*. Boston: Harvard Business School Press.

Cooke, Philip, and Kevin Morgan. 1998. *The Associational Economy: Firms, Regions, and Innovation*. Oxford: Oxford University Press.

CMRG, Creative Management Research Group. 1995. Key Points for New Product Development from Cases of Leading Companies. Research and Development Management, June.

Coase, Ronald H. 1937. The Nature of the Firm. *Econometrica* 4(16): 386–405.

Coase, Ronald H., Oliver E. Williamson, and Sidney G. Winter. 1991. *The Nature of the Firm: Origins, Evolution, and Development*. New York: Oxford University Press.

Cohen, Stephen, Alberto Di Minin, Yasuyuki Motoyama, and Christopher Palmberg. 2009. Persistence of Home Bias for Important R&D in Wireless Telecom and Automobiles. *Review of Policy Research* 26(1–2): 55–76.

Coleman, James Samuel. 1990. *Foundations of Social Theory*. Cambridge, Mass.: Belknap Press of Harvard University Press.

Cooke, Philip. 1998. Introduction: Origins of the Concept. In *Regional Innovation Systems: the Role of Governances in a Globalized World*, edited by H.-J. Braczyk, P. Cooke and M. Heidenreich. London: UCL Press.

Cook, Thomas D., and Donald Thomas Campbell. 1979. *Quasi-experimentation: Design & Analysis Issues for Field Settings*. Chicago: Rand McNally College Pub. Co.

Cooke, Philip, and Kevin Morgan. 1998. *The Associational Economy: Firms, Regions, and Innovation*. Oxford: Oxford University Press.

Cooper, Richard. 2004. A False Alarm: Overcoming Globalization's Discontents. *Foreign Affairs* 83(1): 152–5.

Costanzo, Joseph, Cynthia Davis, Caribert Irazi, Daniel Goodkind, and Roberto Ramirez. 2001. Estimating Unauthorized Migration: 1990 and 2000. In *Demographic Analysis-Population Estimates*. Washington DC: Census Bureau.

Culnan, Mary J., and Lynne M. Markus. 1987. Information technologies. In *Handbook of Organizational Communication: An Interdisciplinary Perspective*, edited by F.M. Jablin. Newbury Park, Calif.: Sage Publications.

Currid, Elizabeth. 2007. *The Warhol Economy: How Fashion, Art, and Music Drive New York City*. Princeton, NJ: Princeton University Press.

Curry, James, and Martin Kenney. 2004. Organizational and Geographic Configuration of the Personal Computer Value Chain. In *Locating Global Advantage: Industry Dynamics in the International Economy*, edited by M. Kenney and R. Florida, Stanford, Calif.: Stanford University Press, 113–41.

Cusumano, Michael A., and Kentaro Nobeoka. 1998. *Thinking Beyond Lean: How Multi-Project Management is Transforming Product Development at Toyota and Other Companies*. New York: Free Press.

Dalton, Donald H., Manuel G. Serapio, and Phyllis Genther Yoshida. 1999. Globalizing Industrial Research and Development. Washington DC: US Department of Commerce, Technology Administration, Office of Technology Policy.

Daniels, P.W. 1993. *Service Industries in the World Economy*, IBG studies in geography. Oxford: Blackwell.

Dobson, Wendy, and Chia Siow Yue. 1997. *Multinationals and East Asian Integration*. Ottawa, Canada: International Development Research Centre.

Dore, Ronald Philip. 1973. *British Factory, Japanese Factory; The Origins of National Diversity in Industrial Relations*. Berkeley: University of California Press.

Doremus, Paul N., William W. Keller, Louis W. Pauly, and Simon Reich. 1998. *Myth of the Global Corporation*. Princeton: Princeton University Press.

Drennan, M.P. 1996. The Dominance of International Finance by London, New York And Tokyo. In *The Global Economy in Transition*, edited by P.W. Daniels and W.F. Lever, Harlow: Longman, 352–71.

Drucker, Peter Ferdinand. 1982. *The New Society: The Anatomy of the Industrial Order*. Westport, Conn.: Greenwood Press.

Dunning, John H. 1983. Changes in the Level and Structure of International Production: The Last One Hundred Years. In *The Growth of International Business*, edited by M. Casson and P.J. Buckley. London: Allen & Unwin.

Dunning, John H. 1992. *Multinational Enterprises and the Global Economy*. Wokingham, England: Addison-Wesley.

Dunning, John H. 2000. *Regions, Globalization, and the Knowledge-Based Economy*. Oxford: Oxford University Press.

Eccles, Robert G., and Dwight B. Crane. 1988. *Doing Deals: Investment Banks at Work*. Boston, Mass.: Harvard Business School Press.

Endo, Ichiro. 1993. Cohabitation of Desired Technology and Skills: A Serendipitous Case in Bubble-Jet Printing Technology. *Technology and Economy* June: 16–23.

———. 1994. Production of Creative Product and Business that Revitalize Firms. Business Research 1.

EPA (Environmental Protection Agency). Global Climate Protection Award 1998. Available from http://www.epa.gov/cppd/awards/climproawards.htm.

Ernst, Dieter. 1994. *Inter-firm Networks and Market Structure: Driving Forces, Barriers and Patterns of Control*. Berkeley, CA: Berkeley Roundtable on International Economy.

Ernst, Dieter. 1997a. From Partial to Systemic Globalization: International Production Networks in the Electronics Industry. In *Working Paper*. Berkeley: Berkeley Roundtable on International Economy (BRIE).

Ernst, Dieter. 1997b. Partners for the China Circle? The East Asian Production Networks of Japanese Electronics Firms. In *The China Circle: Economics and Electronics in the PRC, Taiwan, and Hong Kong*, edited by B. Naughton, Washington, D.C.: Brookings Institution Press, 210–53.

Ernst, Dieter. 2000. Evolutionary Aspects: The Asian Production Networks of Japanese Electronics Firms. In *International Production Networks in Asia*, edited by M. Borrus, D. Ernst and S. Haggard. New York: Routledge, 80–109.

Ernst, Dieter, and Paolo Guerrieri. 1998. International Production Networks and Changing Trade Patterns in East Asia: The Case of the Electronics Industry. *Oxford Development Studies* 26(2): 191–212.

Ernst, Dieter, and John Ravenhill. 2000. Convergence and Diversity: How Globalization Reshapes Asian Production Networks. In *International Production Networks in Asia*, edited by M. Borrus, D. Ernst and S. Haggard. New York: Routledge, 223–53.

Evans, Peter. 2002. The Political Economy of Globalization: Syllabus for Sociology 190.05—Fall. Berkeley, CA.

Feinstein, C. 1990. Britain's Overseas Investments in 1913. *Economic History Review* XLIII(2): 280–95.

Feldman, Maryann P. 1994. *The Geography of Innovation, Economics of Science, Technology, and Innovation*. Boston: Kluwer Academic.

Feldman, Maryann P. 2000. Location and Innovation: The New Economic Geography of Innovation, Spillovers and Agglomeration. In *Oxford Handbook of Economic Geography*, edited by G.L. Clark, M.P. Feldman and M.S. Gertler. Oxford: Oxford University Press, 373–94.

Feldman, Maryann P., and Richard Florida. 1994. Geographic Sources of Innovation: Technological Infrastructure and Product Innovation in the United States. *Annals of Association of American Geographers* 84(2): 210–29.

Ferguson, M. 1992. The Mythology about Globalization. *European Journal of Communication* 7(1): 69–93.

Fields, Gary. 2004. *Territories of Profit: Communications, Capitalist Development, and the Innovative Enterprises of G.F. Swift and Dell Computer*, Innovations and technology in the world economy. Stanford, Calif.: Stanford University Press.

Financial-Times. 2004. Most Respected Companies. *Financial Times*, January 30.

Florida, Richard. 1995. Toward the Learning Region. *Futures* 27(5): 527–36.

Florida, Richard L. 2004. *The Rise of the Creative Class: And How it's Transforming Work, Leisure, Community and Everyday Life*. New York, NY: Basic Books.

Franko, Lawrence G. 1976. *The European Multinationals*. Stamford, CT: Greylock.

Freeman, Christopher. 1982. *The Economics of Industrial Innovation*. 2nd ed. Cambridge, Mass.: MIT Press.

Friedman, Thomas L. 2005. *The World is Flat: A Brief History of the Twenty-First Century*. 1st ed. New York: Farrar, Straus and Giroux.

Friedmann, John. 1986. The World City Hypothesis. *Development and Change* 17: 69–83.

Fröbel, Folker, Jürgen Heinrichs, and Otto Kreye. 1980. *The New International Division of Labour: Structural Unemployment in Industrialised Countries and Industrialisation in Developing Countries*. Cambridge: Cambridge University Press.

Fujimoto, Takahiro. 1999. *The Evolution of a Manufacturing System at Toyota*. New York: Oxford University Press.

Gassmann, Oliver, and Maximilian Von Zedtwitz. 1999. New Concepts and trends in international R&D Organization. *Research Policy* 28: 231–50.

Geddes, Patrick. 1915. *Cities in Evolution; An Introduction to the Town Planning Movement and to the Study of Civics*. London: Williams & Norgate.

Gerlach, Michael L. 1992. *Alliance Capitalism: the Social Organization of Japanese Business*. Berkeley: University of California Press.

Gertler, Meric S. 2003. Tacit Knowledge and Economic Geography of Context, or Undefinable Tacitness of Being (There). *Journal of Economic Geography* 3(1): 75–99.

Gerybadze, A., and Guido Reger. 1999. Globalization of R&D: Recent Changes in the Management of Innovation in Transnational Corporations. *Research Policy* 28: 251–74.

Giddens, Anthony. 2000. *Runaway World: How Globalization is Reshaping Our Lives*. New York: Routledge.

Glaeser, Edward L., Hedi D. Kallal, Jose A. Scheinkman, and Andrei Shleifer. 1992. Growth in Cities. *Journal of Political Economy* 100(6): 1126–52.

Goda. 1999. Product Development Strategy of Sony's "Vaio Note 505" (soni vaio noto 505 no shohin kaihatsu senryaku). Tokyo: JMR Life Research Institute.

Goethe, Johann Wolfgang von. 1808. *Faust*: Part 1.

———. 1832. Faust: Part 2.

Graham, Stephen. 2002. Communication Grids: Cities and Infrastructure. In *Global Networks, Linked Cities*, edited by S. Sassen. New York: Routledge.

Granovetter, Mark. 1985. Economic Action and Social Structure: The Problem of Embeddedness. *American Journal of Sociology* 91: 481–510.

Guillen, Mauro F. 2001. Is Globalization Civilizing, Destructive or Feeble?: A Critique of Five Key Debates in the Social Science Literature. *Annual Review of Sociology* 27(235–60).

Hackman, J. Ricahrd, and C.G. Morris. 1978. Group Tasks, Group Interaction Process, and Group Performance Effectiveness: A Review and Proposed Integration. In *Group processes*, edited by L. Berkowitz, New York: Academic Press, 1–55.

Hajimoto, Yoshioki. 1999. American Way of Venturing (Beikoku ryu bencharingu no jissen shuhou). Tokyo: Chuo Keizai.

Hall, Peter. 2001. Global City-Regions in the Twenty-first Century. In *Global City-Regions*, edited by A.J. Scott. New York: Oxford University Press, 59–77.

———. 2006. The Polycentric Metropolis: Learning from Mega-City Regions in Europe. Paper read at Global Metropolitan Studies Lecture Series, February 13, at Berkeley, CA.

Hall, Peter Geoffrey. 1966. *The World Cities*. New York,: McGraw-Hill.

Hargadon, Andrew B., and Kathleen M. Eisenhardt. 2000. Speed and Quality in New Product Management: An Emergent Perspective on Continuous Organizational Adaptation. In *Quality Movement & Organization Theory*, edited by R.E. Cole and W.R. Scott. London: Sage publications, 331–46.

Harrison, Bennett. 1992. Industrial Districts: Old Wine in New Bottle? *Regional Studies* 26: 469–83.

Harrison, Bennett, Maryellen R. Kelley, and Jon Gant. 1996. Innovative Firm Behavior and Local Milieu: Exploring the Intersection of Agglomeration, Firm Effects and Technological Change. *Economic Geography* 72(3): 233–58.

Harryson, Sigvald. 1998. *Japanese Technology and Innovation Management: From Know-how to Know-who*. Northampton, MA: Edward Elgar.

Harvey, D. 1989. *The Condition of Postmodernity: An Enquiry into the Origins of Cultural Change*. Oxford: Basil Blackwell.

Held, David, Anthony McGrew, David Goldblatt, and Jonathan Perraton. 1999. *Global Transformations: Politics, Economics, and Culture*. Stanford: Stanford University Press.

Helper, Susan, John Paul MacDuffie, and Charles F. Sabel. 2000. Pragmatic Collaborations: Advancing Knowledge while Controlling Opportunism. *Industrial and Corporate Change* 9(3): 443–89.

Henderson, Rebecca, and Kim B. Clark. 1990. Architectural Innovation: The Reconfiguration of Existing Product Technology and the Failure of Established Firms. *Administrative Science Quarterly* 35(1): 9–30.

Hino, Satoshi. 2002. A Study of Toyota Management System: Principles for Continuous Growth (toyota keiei sisutemu no kenkyu: eizokuteki seicho no genri). Tokyo: Diamond Inc.

Hirschman, Charles, Philip Kasinitz, and Josh DeWing. 1999. *The Handbook of International Migration: The American Experience*. New York: Russel Sage Foundation.

Hirst, Paul, and Graham Thompson. 1999. *Globalization in Question: The International Economy and the Possibilities of Governance*. 2nd ed. Cambridge: Polity and Oxford University Press.

Hirst, Paul, and Graham Thompson. 2003. The Limits to Economic Globalization. In *The Handbook of Globalisation*, edited by J. Michie. Cheltenham, U.K.: Edward Elgar, 335–48.

Hodder, B.W., and Roger Lee. 1974. *Economic Geography*. London: Methuen.

Honda, Inc. 2005. Operations overview. Honda 2005 [cited May 5, 2005]. Available from http://corporate.honda.com/america/overview.aspx.

Hoshi, Takeo. 1994. Economic Role of Corporate Grouping and the Main Bank System. Edited by M. Aoki and R.P. Dore, *The Japanese firm: the sources of competitive strength*. Oxford: Oxford University Press, 385-309.

Hounshell, David A., and John K. Smith. 1988. *Science and Corporate Strategy: Du Pont R&D, 1902–1980, Studies in Economic History and Policy*. Cambridge: Cambridge University Press.

HP (Hewlett-Packard). 2006. HP Labs: Worldwide Site 2006 [cited March 1, 2006]. Available from http://www.hpl.hp.com/about/sites.html.

Hymer, Stephen. 1972. Multinational Corporation and the Law of Uneven Development. In *Economics and World Order from the 1970s to the 1990s*, edited by J. Bhagwati. London: Collier-Macmillan, 113–40.

Ikari, Yoshiro. 1999. *Era of Hybrid Car: A Story of Toyota "Prius" Development (haiburiddoka no jidai)*. Tokyo: Kojinsha.

Imai, Kenichi, and Ryutaro Komiya. 1989. *Nihon no kigyo. Shohan*. ed. Tokyo: Tokyo Daigaku Shuppankai.

Imai, Kenichi, Ryutaro Komiya, Ronald Philip Dore, and D.H. Whittaker. 1994. *Business enterprise in Japan: views of leading Japanese economists*. Cambridge, Mass.: MIT Press.

Intel. 2002. 10-K Filing. Santa Clara, CA: Intel Corporation.

Irwin, Douglas A., and Peter J. Klenow. 1994. Learning-by-Doing Spillovers in the Semiconductor Industry. *Journal of Political Economy* 102(6): 1200–27.

Itami, Hiroyuki, and Thomas W. Roehl. 1987. *Mobilizing Invisible Assets*. Cambridge, Mass.: Harvard University Press.

Itazaki, Eishi. 1999. *Innovation of Toyota's Car: The Prius that Shocked the Entire World (kakushin toyota jidousha: sekai wo sinkan saseta puriusuno shougeki)*. Tokyo: Nikkann Kogyo Shimpo.

Jaffe, Adam B., Manuel Trajtenberg, and Rebecca Henderson. 1993. Geographic Location of Knowledge Spillovers as Evidenced by Patent Citations. *Quarterly Journal of Economics* 63(3): 577–98.

Jaffe, Adam B., and Manuel Trajtenberg. 2002. *Patents, Citations, and Innovations: A Window on the Knowledge Economy.* Cambridge, Mass.: MIT Press.

JCOTY, Japan Car of the Year Committee. 2005. Japan Car of the Year History 1997 [cited September 17, 2005]. Available from http://www.jcoty.org/history/1997.html.

Johnson, Chalmers A. 1982. *MITI and the Japanese Miracle: the Growth of Industrial Policy, 1925–1975.* Stanford, Calif.: Stanford University Press.

Johnston, Robert E. 1966. Technical Progress and Innovation. *Oxford Economic Papers* 18(2): 158–76.

Jones, Geoffrey. 2005. *Multinationals and Global Capitalism: From the Nineteenth to the Twenty-First Century.* Oxford: Oxford University Press.

Jovanovic, Boyan, and Yaw Nyarko. 1996. Learning by Doing and the Choice of Technology. *Econometrica* 64(6): 1299–310.

Jones, Geoffrey. 2005. *Multinationals and Global Capitalism: From the Nineteenth to the Twenty-First Century.* Oxford: Oxford University Press.

JPO (Japan Patent Office). 2004. Report of Technological Status of Patent Filing. Tokyo.

Katzenstein, Peter. 2003. Japan, Technology and Asian regionalism in Comparative Perspective. In *The Resurgence of East Asia: 500, 150 and 50 Year Perspectives*, edited by G. Arrighi, T. Hamashita, and M. Selden. London: Routledge, 214–58.

Kawaguchi, Yoji. 2003. *Strategy of Sony's Contents (soni no contentsu senryaku).* Tokyo: Japana Mangement Association of Manufacturers.

Keeling, D.J. 1995. Transportation and the world city paradigm. In *World Cities in a World-system*, edited by P.L. Knox and P.J. Taylor. Cambridge: Cambridge University Press, 115–31.

Keller, William W., and Richard J. Samuels. 2003. *Crisis and Innovation in Asian Technology.* Cambridge: Cambridge University Press.

Kenney, Martin. 2000. *Understanding Silicon Valley: The Anatomy of an Entrepreneurial Region.* Palo Alto: Stanford University Press.

Keohane, Robert O., and Joseph S. Nye. 2000. Globalization: What's New? What's Not? (And So What?). *Foreign Policy* 118: 104–19.

Kogut, Bruce Mitchel. 2004. Conclusion: From Regions and Firms to Multinational Highways: Knowledge and Its Diffusion as a Factor in the Globalization of Industries. Edited by M. Kenney and R.L. Florida, *Locating Global Advantage: Industry Dynamics in the International Economy.* Stanford, Calif.: Stanford University Press, 261–82.

Kohno, Masaru. 2003. A Changing Ministry of International Trade and Industry. In *Japanese Governance: Beyond Japan Inc*, edited by J.A. Amyx and P. Drysdale. London: RoutledgeCurzon, 93–112.

Kondratiev, Nikolai D. 1924. Long Waves in Economic Life. *Reviews of Economic Statistics* 17: 105–15.

Koyanagawa, Takashi, Michinobu Inoue, and Kazuya Shibasaki. 2000. Thermal Design Technology for Small and Slim Notebook PCs. *Toshiba Review* 55(4): 9–12.

Kozato, Yasuo. 1999. Canon: R&D as the Motivating Force for Continuous Growth and Diversification. In *Managing Global Innovation: Uncovering the Secrets of Future Competitiveness*, edited by R. Boutellier, O. Gassmann and m. Von Zedtwitz. Berlin: Springer, 316–33.

Kresel, P.K. 1995. The Determinants of Urban Competitiveness: A Survey. In *North American Cities and the Global Economy*, edited by P.K. Kresel and G. Gaert. London: Sage, 45–68.

Krugman, Paul. 1991. *Geography and Trade*. Cambridge MA: MIT Press.

Kunzmann, K.R. 1998. World City Regions in Europe: Structural Change and Future Challenges. Edited by F.-c. Lo and Y.-m. Yeung, *Globalization and the World of Large Cities*. Tokyo: United Nations University Press, 37–76.

Kuznets, Simon Smith. 1930. *Secular Movements in Production and Prices; Their Nature and Their Bearing upon Cyclical Fluctuations*. Boston: Houghton Mifflin company.

Kuznets, Simon Smith, and Ernest Rubin. 1954. Immigration and the Foreign-Born. In *Occasional Paper*. Boston: National Bureau of Economic Research.

Lakoff, George, and Mark Johnson. 1980. *Metaphors we Live by*. Chicago: University of Chicago Press.

Lam, Alice. 2003. Organisational Learning in Multinationals: R&D Networks of Japanese and US MNEs in the U.K. *Journal of Management Studies* 40(3): 673–703.

Lazonick, William. 1991. *Business Organization and the Myth of the Market Economy*. Cambridge: Cambridge University Press.

Le, Hue P. 1998. Progress and Trend in Ink-jet Printing Technologies. *Journal of Imaging Science and Technology* 42(1): 49–62.

Lee, Roger, and Jane Wills. 1997. *Geographies of Economies*. London: Arnold.

Lester, Richard K., and Michael J. Piore. 2004. *Innovation, the Missing Dimension*. Cambridge, Mass.: Harvard University Press.

Liker, Jeffrey K. 2004. *Toyota Way: 14 Management Principles from the World's Greatest Manufacturer*. New York: McGraw-Hill.

Llewelyn-Davies, University College London. Bartlett School of Planning, and Comedia. 1996. Four World Cities: A Comparative Study of London, Paris, New York, and Tokyo: Summary Report for Department of the Environment & Government Office for London. London: Llewelyn-Davies.

Lundvall, B.A., and B. Johnson. 1994. The Learning Economy. *Journal of Industry Studies* 1: 23–42.

Lyons, D., and S. Salmon. 1995. World Cities, Multinational Corporations, and Urban Hierarchy: The Case of the United States. In *World Cities in a World*

System, edited by P.L. Knox and P.J. Taylor. Cambridge: Cambridge University Press, 98–114.

Macher, Jeffrey T., and David C. Mowery. 2008. *Innovation in Global Industries: U.S. Firms Competing in a New World*. Washington, D.C.: National Academies Press.

Maddison, A. 1987. Growth and Slow-Down in Advanced Capitalist Economies: Techniques of Quantitative Assessments. *Journal of Economic Literature* XXV(2): 649–98.

Malmberg, Anders, and Peter Maskell. 1997. Toward an Explanation of Regional Specialization and Industry Agglomeration. *European Planning Studies* 5(1): 25–41.

Marek, S. 1992. Private networks directory. Satellite Communications.

Markusen, Ann Roell. 1985. *Profit Cycles, Oligopoly and Regional Development*. Cambridge MA: MIT Press.

———. 1994. Studying Regions by Studying Firms. *Professional Geographer* 46(4): 477–90.

———. 1996. Sticky Places in Slippery Space: A Typology of Industrial Districts. *Economic Geography* 72(3): 293–313.

———. 2003. Fuzzy Concepts, Scanty Evidence, Policy Distance: The Case for Rigour and Policy Relevance in Critical Regional Studies. *Regional Studies* 37(6&7): 701–17.

Marshall, Alfred. 1898. *Principles of Economics*. 4th ed. 6 vols. London: Macmillan and Company.

Maskell, Peter. 2001. Firm in Economic Geography. *Economic Geography* 77(4): 329–43.

Massey, Doreen. 1979. *Spatial Divisions of Labour: Social Structures and the Geography of Production*. London: Macmillan.

McClelland, Peter, and Richard Zeckhauser. 2004. *Demographic Dimensions of the New Republic: American Interregional Migration, Vital Statistics, and Manumissions, 1800–1860*. 1st paperback ed. New York: Cambridge University Press.

McKendrick, David G., Richard F. Doner, and Stephan Haggard. 2000. *From Silicon Valley to Singapore: Location and Competitive Advantage in the Hard Disk Drive Industry*. Palo Alto, CA: Stanford University Press.

McWilliams, Gary. 1997. If at First You Falter, Reboot; Japan Readies a New Assault on America's PC Markets. *BusinessWeek*, June 30, 81.

MEXT, Ministry of Education, Culture, Sports, Science and Technology. 2004. White Paper on Science and Technology. Tokyo: MEXT.

Microsoft, Inc. 2008. Microsoft Research Home 2008 [cited September 15, 2008]. Available from http://research.microsoft.com/.

Miller, Roger. 1994. Global R&D Networks and Large-Scale innovations: The Case of the Automobile Industry. *Research Policy* 23(1): 27–46.

Miyazaki, Masaya. 2002. Inkjet Printer Industry's Development Process: 1977–1997. *Akamon Management Review* 1(2): 159–97.

Mowery, David C. and Nathan Rosenberg. 1998. *Paths of innovation: technological change in 20th century America.* Cambridge, U.K.: Cambridge University Press.

New York Times. 2005. Shakeup at Sony Puts Westerner in Leader's Role. *New York Times*, March 7.

Nijman, J. 1996. Breaking the Rules: Miami in the Urban Hierarchy. *Urban Geography* 17: 5–22.

Nikkan, Kogyo Shimbun. 2003. *Facts of Canon.* Tokyo.

Nikkei. 2004. Nikkei Prism. Nihon Keizai Shimbun, April 1.

Nikkei (Nihon Keizai Shinbun). 1997. My Resume (Series). Nikkei Newspaper, March 14.

———. 2001. Canon: Secrets for Its Revival of High Profits. Tokyo: Nikkei, Nihon Keizai Shinbun.

———. 2005. Official Appointment of Mr. Mitarai by Chairman Okuda. Nikkei Newspaper, October 17.

Ministry of Finance. 2002. Report on the Studying Group of Industrial Hollowing-Out and Tariffs: Win-Win Scenario Beyond Hollowing-Out. Tokyo: Ministry of Finance.

Nobeoka, Kentaro, and Takahiro Fujimoto. 2004. Organizational capacity of product development: International competitiveness of Japanese automobile firms. Tokyo: National Institute of Advanced Industrial Society and Technology (AIST).

Nohria, Nitin, and Robert G. Eccles. 1992. Face-to-Face: Making Network Organizations Work. In *Networks and Organizations: Structure, Form, and Action*, edited by N. Nohria and R.G. Eccles. Boston Mass.: Harvard Business School Press, 388-308.

Nohria, Nitin, and Sumantra Ghoshal. 1997. *Differentiated Network: Organizing Multinational Corporations for Value Creation.* San Francisco: Jossey-Bass Publishers.

Nonaka, Ikujiro, and Hirotaka Takeuchi. 1995. *The Knowledge-Creating Company: How Japanese Companies Create the Dynamics of Innovation.* New York: Oxford University Press.

Nonaka, Ikujiro, and Hirotaka Takeuchi. 2000. *The Knowledge Creating Company.* In Japanese, 11th ed. Tokyo: Toyo Keizai Shimpo.

Nooteboom, B. 1999. Innovation, Learning and Industrial Organisation. *Cambridge Journal of Economics* 23: 127–50.

O'Brien, Richard. 1992. *Global Financial Integration: The End of Geography.* New York: Council on Foreign Relations Press.

O'Connor, K. 1995. Change in the Pattern of Airline Services and City Development. Edited by J.F. Brotchie, *Cities in Competition: Productive and Sustainable Cities for the 21st Century.* Melbourne: Longman Australia.

OECD. 1993a. Globalization in the Pharmaceutical Industry. Paris.

————. 1993b. Globalization of Industrial Activities: Sector Case Study of Globalization in the Computer Industry. Paris: Organization for Economic Cooperation and Development.

————. 2005a. Handbook on Economic Globalisation Indicators. Paris: Organization for Economic Cooperation and Development.

————. 2005b. Measuring Globalisation: Economic Globalisation Indicators. Paris: Organization for Economic Cooperation and Development.

OECD. 2009. Beyond 20/20. Online Database.

Ohmae, Kenichi. 1995. *The End of the Nation State: The Rise of Regional Economies*. New York: Free Press.

Okuda, Hiroshi. 2004. We Can and We Will: Presentation to Investors. Tokyo.

Ono, Taiichi. 1978. Toyota Production System: Toward Management Without Economies of Scale (toyota prodakushon sisutemu: datsu kibo no keiei wo mezashite). Tokyo: Diamond.

————. 2001. Shopfloor Management (genba no keiei). Tokyo: JSPM Management Center.

Paci, Raffaele, and Stefano Usai. 2000. Technological Enclaves and Industrial Districts: Analysis of the Regional Distribution of Innovative Activity in Europe. *Regional Studies* 34(2): 97–114.

Panasonic, EV Synergy. 2005. *Company Profile* 2005 [cited August 9, 2005]. Available from http://www.peve.panasonic.co.jp/profile.html.

Passel, Jeffrey, and D'Vera Cohn. 2008. Trends in Unauthorized Immigration: Undocumented Inflow Now Trails Legal Inflow. In *Pew Research Project*. Washington DC: Pew Hispanic Center.

Patel, P., and Keith Pavitt. 1991. Large Firms in the Production of the World's Technology: An Important Case of "Non-Globalization". *Journal of International Business Studies* 22(1): 1–21.

Patton, Donald, and Martin Kenney. 2003. Spatial Distribution of Entrepreneurial Support Networks: Evidence from Semiconductor Initial Public Offerings from 1996 through 2000. Berkeley: BRIE.

Pavitt, Keith. 1984. Sectoral Patterns of Technical Change: Toward a Taxonomy and a Theory. *Research Policy* 13(6): 343–73.

Pearce, Robert D., and Marina Papanastassiou. 1996. R&D Networks And Innovation: Decentralised Product Development in Multinational Enterprises. *R&D Management* 26(4): 315–33.

Perroux, Francois. 1955. A Note on the Notion of Growth Pole. Economie Appliquee.

Piore, Michael J., and Charles F. Sabel. 1984. Second Industrial Divide: Possibilities For Prosperity. New York: Basic Books.

Platt, D.C.M. 1980. British Portfolio Investment before 1870: Some Doubts. *Economic History Review* XXXIII(1): 1–16.

Polanyi, Michael. 1966. *Tacit Dimension*. London: Routledge & Kegan Paul.

Porter, Michael E. 1985. *Competitive Advantage: Creating and Sustaining Superior Performance*. New York: Free Press, Collier Macmillan.

————. 1990. *The Competitive Advantage of Nations*. New York: Free Press.

————. 1994. The Role of Location in Competition. *Journal of the Economics of Business* 1(1): 35–9.

————. 1995. The Competitive Advantage of the Inner City. *Harvard Business Review* (May–June): 55–71.

————. 1998a. *The Competitive Advantage of Nations: With a New Introduction*. New York: Free Press.

————. 1998b. Clusters and the New Economics of Competition. *Harvard Business Review* 76(6): 77–90.

————. 2000. Location, Competition, and Economic Development: Local Clusters in a Global Economy. *Economic Development Quarterly* 14(1): 15–34.

Prakash, Om. 1985. *The Dutch East India Company and the economy of Bengal, 1630–1720*. Princeton N.J.: Princeton University Press.

Prestowitz, Clyde V. 1989. *Trading Places: How We Are Giving Our Future to Japan and How to Reclaim It*. New York: Basic Books.

Putnam, Robert D., Robert Leonardi, and Raffaella Nanetti. 1993. *Making Democracy Work: Civic Traditions in Modern Italy*. Princeton, N.J.: Princeton University Press.

Ratanawaraha, Apiwat, and Karen R. Polenske. 2007. *Measuring the Geography of Innovation: A Literature Review*. Edited by K.R. Polenske, *Economic Geography of Innovation*. Cambridge: Cambridge University Press, 30–59.

Ravenhill, John. 1999. Japanese and US Subsidiaries in Asia: Host Country Effects. In *Japanese Multinationals in Asia: Regional Operations in Comparative Perspective*, edited by D.J. Encarnation. Oxford: Oxford University Press, 261–84.

Reddy, Prasada. 2000. *The Globalization of Corporate R&D: Implications for Innovation Systems in Host Countries*. London: Routledge.

Reingold, Edwin. 1999. *Toyota: People, Ideas and the Challenge of the New*. London: Penguin Books.

Rimmer, Peter J. 1998. Transport and Telecommunications among World Cities. Edited by F.-c. Lo and Y.-m. Yeung, *Globalization and the World of Large Cities*. Tokyo: United Nations University Press, 433–70.

RJC (Automotive Researchers' and Journalists' Conference of Japan). History of Year Prize (iya sho no rekishi) 1998 [cited. Available from http://www.rjc. or.jp/prize/year1998/rjc98.html.

Rosenberg, Nathan. 1982. *Inside the Black Box: Technology and Economics*. Cambridge: Cambridge University Press.

————. 1994. *Exploring the Black Box: Technology, Economics, and History*. Cambridge: Cambridge University Press.

Sabel, Charles F. 1993. Studied Trust: Building New Forms of Cooperation in a Volatile Economy. In *Explorations in Economic Sociology*, edited by R. Swedberg. New York: Russell Sage Foundation, 104–44.

Sachs, Jeffrey, and Andrew Warner. 1995. Economic Reform and the Process of Global Integration. In *Brookings Papers on Economic Activity*. Washington DC: Brookings Institute.

Sakakibara, Kinonori, and Youichi Matsumoto. 2004. Possessive Possibility of Innovations: A Case of Canon: National Institute of Advanced Industrial Science and Technology.

Sakakibara, Kinonori, and D.E. Westney. 1992. Japan's Management of Global Innovation: Technology Management Crossing Borders. In *Technology and the Wealth of Nations*, edited by N. Rosenberg, R. Landau and D.C. Mowery. Stanford, Calif.: Stanford University Press, 327–43.

Samuels, Richard J. 1994. *"Rich Nation, Strong Army": National Security and the Technological Transformation of Japan, Cornell Studies in Political Economy*. Ithaca: Cornell University Press.

Sassen, Saskia. 1991. *The Global City*: New York, London, Tokyo. Princeton: Princeton University Press.

———. 2001. Global Cities and Global City-Regions: A Comparison. In *Global City-Regions*, edited by A.J. Scott. Oxford: Oxford University Press, 78–95.

Savitch, H.V., and P. Kantor. 1995. City Business: An International Perspective on Marketplace Politics. *International Review of Urban and Regional Research* 19: 495–512.

Saxenian, AnnaLee. 1994. *Regional Advantage: Culture and Competition in Silicon Valley and Route 128*. Cambridge, Mass.: Harvard University Press.

Saxenian, AnnaLee. 2006. *The New Argonauts: Regional Advantage in a Global Economy*. Cambridge, MA: Harvard University Press.

Saxenian, AnnaLee, Yasuyuki Motoyama, and Xiaohong Quan. 2002. *Local and Global Networks of Immigrant Professionals in Silicon Valley*. San Francisco, CA: Public Policy Institute of California.

Schegloff, Emanuel A. 1987. Between Micro and Macro: Contexts and Other Connections. In *The Micro-macro Link*, edited by J.C. Alexander. Berkeley: University of California Press.

Schmidt, Vivien. 1995. The New World Order, Incorporated: The Rise of Business and the Decline of the Nation-State. *Daedalus* 124(2): 75–106.

Schoenberger, Erica J. 1997. *Cultural Crisis of the Firm*. Cambridge Mass.: Blackwell.

Schumpeter, Joseph Alois. 1912. Theorie der wirtschaftlichen entwicklung (Theory of Economic Development). Leipzig: Duncker & Humblot.

Schumpeter, Joseph Alois. 1934. *The Theory of Economic Development; An Inquiry into Profits, Capital, Credit, Interest, and the Business Cycle, Harvard Economic Studies. vol. XLVI*. Cambridge, Mass.: Harvard University Press.

———. 1939. *Business Cycles: A Theoretical, Historical, and Statistical Analysis of the Capitalist Process*. 1st ed. New York: McGraw-Hill Book Company, Inc.

———. 1943. *Capitalism, Socialism, and Democracy*. London: G. Allen & Unwin ltd.

———. 1947. Creative Response in Economic History. *Journal of Economic History* 7(2): 149–59.

———. 1954. *History of Economic Analysis*. New York: Oxford University Press.

———. 1983. *The Theory of Economic Development: An Inquiry into Profits, Capital, Credit, Interest, and the Business Cycle*, Social science classics series. New Brunswick, N.J.: Transaction Books.

Scott, Alan J. 1998. *Regions and the World Economy: The Coming Shape of Global Production, Competition, and Political Order*. Oxford: Oxford University Press.

Segal, Aaron, Patricia M. Chalk, and J. Gordon Shields. 1993. *An Atlas of International Migration*. London: Hans Zell.

Seki, Mitsuhiro. 1999. The Destruction of the Full-Set Industrial Structure—East Asia's Tripolar Structure. In *The Transformation of the Japanese Economy*, edited by K. Sato. Armonk, N.Y.: M.E. Sharpe, 321–39.

Serow, William J., Charles B. Nam, David F. Sly, and Robert H. Weller. 1990. *Handbook on International Migration*. Westport, CT: Greenwood Press.

Sobek, Durward, Allen C. Ward, and Jeffrey K. Liker. 1999. Toyota's Principles of Set-Based Concurrent Engineering. *Sloan Management Review* 40(2): 67–83.

Solow, Robert. 1957. Technical Change and the Aggregate Production Function. *Review of Economics and Statistics* 39: 312–20.

Sony. 1997. *Vaio 505 Products* [cited March 21, 2005]. Available from http://www.sony.jp/ProductsPark/Consumer/PCOM/PCG-707.html.

———. 2000. Annual Report. Tokyo: Sony Inc.

———. 2001. Sony Autobiography (soni jijyo den). Tokyo: Sony Public Relations Division.

———. 2002. Annual Report. Tokyo, Japan.

———. 2004a. Interviewing development engineers 2003 [cited May 30, 2004]. Available from http://www.vaio.sony.co.jp/Products/Inside/X505/index.html.

———. 2004b. Annual Report. Tokyo: Sony Inc.

———. 2004c. Social and Environmental Report: Sony, Inc.

———. 2004d. Vaio PCG-505 2004 [cited May 30, 2004]. Available from http://www.sony.jp/ProductsPark/Consumer/PCOM/VAIO/Note505/index.html.

———. 2005a. Vaio 505 Products 1997 [cited March 21, 2005]. Available from http://www.sony.jp/ProductsPark/Consumer/PCOM/PCG-707.html.

———. 2005b. Interviewing Development Engineers. Sony, Inc. 2005 [cited April 6, 2005]. Available from http://www.vaio.sony.co.jp/Products/Inside/R505/r01.html.

———. 2005c. Organizational Structure 2005 [cited March 18, 2005]. Available from http://www.sony.co.jp/SonyInfo/CorporateInfo/Data/organization.html.

———. 2005d. Sony Corporation Announces New Management Structure. Tokyo.

———. 2006. Annual Report. Tokyo: Sony Inc.

Sony-EMCS. 2004a. *Work Style Interview (a)*. Sony EMCS 2004 [cited 6/30 2004]. Available from http://www.sonyemcs.net/nagano_tec/workstyle/w_05.html.

————. 2004b. *Work Style Interview (b)* 2004 [cited March 10, 2004]. Available from http://www.sonyemcs.net/nagano_tec/workstyle/w_02.html.

————. 2005. History of Sony EMCS (soni emcs no keii) [cited February 7, 2005. Available from http://www.sonyemcs.net/nagano_tec/company/c_08.html.

Sony-Marketing. 2005. Corporate Profile. Sony-Marketing 2004 [cited April 14, 2005]. Available from http://www.sony.jp/CorporateCruise/SMOJ-info/Profile.html.

Sproull, Lee, and Sara Kiesler. 1986. Reducing Social Context Cues: Electronic Mail in Organizational Communication. *Management Science* 32(11): 1492–512.

————. 1991. Computers, Networks, and Work. *Scientific American* 265(3): 116–27.

Strange, Susan. 1995. The Defective State. *Daedalus* 124(2): 55–74.

Stiftel, Bruce, Deden Rukmana, and Bhuiyan Alam. 2004. Faculty Quality at U.S. Graduate Planning Schools: A National Research Council-Style Study. *Journal of Planning Education and Research* 24: 6–22.

Storper, Michael. 1995. The Resurgence of Regional Economies Ten Years Later: The Region as a Nexus of Untraded Interdependencies. *European Urban and Regional Studies* 2: 191–221.

Storper, Michael. 1997. *The Regional World: Territorial Development in a Global Economy, Perspectives on economic change*. New York: Guilford Press.

Storper, Michael, and Anthony Venables. 2004. Buzz: Face-To-Face Contact and the Urban Economy. *Journal of Economic Geography* 4(4): 251–70.

Storper, Michael, and Richard Walker. 1989. *The Capitalist Imperative: Territory, Technology, and Industrial Growth*. Oxford: Blackwell.

Studer-Noguez, Isabel. 2002. *Ford and the Global Strategies of Multinationals: The North American Auto Industry*, Routledge studies in international business and the world economy. London: Routledge.

Sturgeon, Timothy J. 1997. Turnkey Production Networks: A New American Model of Industrial Organization? Berkeley: Berkeley Roundtable on International Economy.

Sturgeon, Timothy J., and Richard Florida. 2004. Globalization, Deverticalization, and Employment in the Motor Vehicle Industry. In *Locating Global Advantage: Industry Dynamics in the International Economy*, edited by M. Kenney and R.L. Florida. Stanford, Calif.: Stanford University Press, 52–81.

Sturgeon, Timothy J., Johannes Van Biesebroeck, and Gary Gereffi. 2008. Value Chains, Networks and Clusters: Reframing the Global Automotive Industry. *Journal of Economic Geography* 8: 297–321.

Tateishi, Yasunori. 2002. Sony Revolution (soni kakumei). Tokyo: President.

Taylor, Michael, and Bjorn Asheim. 2001. Concept of the Firm in Economic Geography. *Economic Geography* 77(4): 315–28.

Taylor, Frederick Winslow. 1903. Shop Management: A paper read before the American Society of Mechanical Engineers. New York.

Taylor, Frederick Winslow. 1911. *The Principles of Scientific Management*. London: Harper & Brothers.

Taylor, Peter J. 1997. Hierarchical Tendencies amongst World Cites: A Global Research Proposal. *Cities* 14: 323–32.

———. 1999. So-Called "World Cities": The Evidential Structure within a Literature. *Environment and Planning A* 31(11).

Taylor, Peter J., and Robert E. Lang. 2005. *U.S. Cities in the "World City Network"*. In Metropolitan Policy Program: Survey Series. Washington DC: Brookings Institution.

Taylor, Peter J., D.R.F. Walker, and J.V. Beaverstock. 2002. Firms and Their Global Service Networks. In *Global Networks, Linked Cities*, edited by S. Sassen. New York: Routledge, 93–116.

Teece, David J. 1996. Firm Organization, Industrial Structure, and Technological Innovation. *Journal of Economic Behavior and Organization* 31: 193–224.

Toyota, Central R&D Lab. 2005. Overview of the Company (kaisha gaiyou) 2005 [cited May 12, 2005]. Available from http://www.tytlabs.co.jp/japanese/comp/outline01.html.

Toyota, Inc. 1994. Outline of Technical Center. Toyoda, Japan.

———. Press Release 1997. Available from www.toyota.co.jp/press.

———. 2000. Special Story: Creation of Prius. Toyota: Toyota.

———. 2002. Global Vision 2010. Toyota, Japan.

———. 2003. Annual Report. Tokyo, Japan.

———. 2004a. Annual Report. Toyota, Japan.

———. 2004b. Overview of Toyota: data of worldly Toyota (toyota no gaikyo: suuji de miru sekai no nakano toyota). Toyoda: Toyota, Inc.

———. 2004c. Technology Report: Toyota Hybrid System 2004 [cited October 13, 2004]. Available from www.toyota.co.jp/jp/tech/environment/ths2/seisan.html.

———. 2005a. History of Toyota [cited April 26, 2005]. Available from http://www.toyota.co.jp/jp/about_toyota/history/index.html.

———. 2005b. Company Profile [cited August 15, 2005]. Available from http://www.toyota.co.jp/en/about_toyota/manufacturing/.

———. 2005c. Manufacturing: Worldwide operations. Toyota [cited May 5, 2005]. Available from http://www.toyota.co.jp/en/about_toyota/manufacturing/worldwide.html.

———. 2006. Prius Key Features 2006 [cited January 23, 2006]. Available from http://www.toyota.com/vehicles/2005/prius/key_features/hybrid_syn_drive.html.

Tyson, Laura D'Andrea. 1993. *Who's Bashing Whom?: Trade Conflicts in High-Technology Industries*. Washington, DC: Institute for International Economics.

UNCTAD. 1994. *World Investment Report*. New York: United Nations Conference on Trade and Development.

———. 1997. *World Investment Report*. New York: United Nations Conference on Trade and Development.

————. 1998. *World Investment Report*. New York: United Nations Conference on Trade and Development.

Urata, Shujiro. 1999. Intrafirm Technology Transfer by Japanese Multinationals in Asia. In *Japanese Multinationals in Asia: Regional Operations in Comparative Perspective*, edited by D.J. Encarnation. New York: Oxford University Press, 143–62.

USPTO. 2006. Preliminary List of Top Patenting Organizations During the Calendar Year 2005 [cited January 2, 2006]. Available from http://www.uspto. gov/web/offices/ac/ido/oeip/taf/reports_top10.htm#TOP_10.

Vernon, Raymond. 1966. International Investment and International Trade in the Product Cycle. *Quarterly Journal of Economics* 80: 190–207.

Vogel, Steven Kent. 2006. *Japan Remodeled: How Government and Industry are Reforming Japanese Capitalism*. Edited by P. Katzenstein, *Cornell Studies in Political Economy*. Ithaca, N.Y.: Cornell University Press.

Ward, Allen, Jeffrey Liker, John Cristiano, and Durward Sboek. 1995. Second Toyota Paradox: How Delaying Decisions Can Make Better Cars Faster. *Sloan Management Review* 36(3): 43–61.

Warf, Barney. 1995. Telecommunications and the Changing Geographies of Knowledge Transmission in the Late 20th Century. *Urban Studies* 32(2): 361–78.

Westney, D.E. 1999. Changing Perspectives on the Organization of Japanese Multinational Companies. In *Japanese Multinationals Abroad: Individual and Organizational Learning*, edited by S. Beechler and A. Bird. New York: Oxford University Press, 11–32.

White, Harry Dexter. 1933. *The French International Accounts: 1880–1913, Harvard Economic Studies*. Cambridge, MA: Harvard University Press.

Whitehead, Alfred North. 1930. *Science and the Modern World*, by Alfred North Whitehead, Lowell lectures, 1925. Cambridge: The University press.

Wilkins, Mira. 1977. Modern European Economic history and the Multinationals. *Journal of European Economic History* 6: 575–95.

Wilkins, Mira. 1991. *The Growth of Multinationals*. Aldershot Hants, England: Edward Elgar.

Williamson, Oliver E. 1975. *Markets and Hierarchies, Analysis and Antitrust Implications: A Study in the Economics of Internal Organization*. New York: Free Press.

————. 1985. *The Economic Institutions of Capitalism: Firms, Markets, Relational Contracting*. New York: Free Press Collier Macmillan.

Wilson, Charles. 1976. The Multinational in Historical Perspective. In *Strategy and Structure of Big Business*, edited by K. Nakagawa. Tokyo: University of Tokyo Press.

Witt, Michael A. 2006. *Changing Japanese Capitalism: Societal Coordination and Institutional Adjustment*. Cambridge: Cambridge University Press.

Womack, James P., and Daniel T. Jones. 1996. *Lean Thinking: Banish Waste and Create Wealth in your Corporation*. New York, NY: Simon & Schuster.

Womack, James P., Daniel T. Jones, and Daniel Roos. 1991. *The Machine That Changed the World: How Japan's Secret Weapon in the Global Auto Wars Will Revolutionize Western Industry*. 1st Harper perennial ed. New York, NY: HarperPerennial.

Yin, Robert K. 1994. *Case Study Research: Design and Methods*. 2nd ed., Applied social research methods series; vol. 5. Thousand Oaks: Sage Publications.

————. 2003. *Case Study Research: Design and Methods*. 3rd ed., Applied social research methods series; vol. 5. Thousand Oaks, Calif.: Sage Publications.

Yoneyama, Shigemi. 1996. Transformative Capability as a Source of Sustainable Competitive Advantages: A Case Study of Canon's Printer Technology Development. *Commercial Review of Seinan Gakuin University* 43(1): 105–68.

Young, Alwyn. 1991. Invention and Bounded Learning by Doing. NBER Working Paper W3712.

Zeile, William. 1997. U.S. Intrafirm Trade in Goods. In *Survey of Current Business*. Washington DC: Bureau of Economic Analysis.

Zevin, Robert. 1992. Are World Financial Markets More Open? If So, Why and With What Effects? In *Financial Openness and National Autonomy*, edited by T. Banuri and J.B. Schor. Oxford: Clarendon Press, 43–83.

Zook, Matthew. 2004. Knowledge Brokers: Venture Capitalists, Tacit Knowledge and Regional Development. *International Journal of Urban and Regional Research* 28(3): 621–41.

Zucker, Lynne G., and Michael R. Darby. 1996. Star Scientists and Institutional Transformation: Patterns of Invention and Innovation in the Formation of the Biotechnology Industry. Paper read at Proceedings to National Academy of Science, October, at Irvine, CA.

Zucker, Lynne G., Michael R. Darby, and Jeff Armstrong. 1998. Geographically Localized Knowledge: Spillovers or Markets? *Economic Inquiry* 36(1): 65–86.

Index

Figures and tables indexed in bold